A Yankee on Puget Sound

Pioneer Dispatches of Edward Jay Allen, 1852–1855

Karen L. Johnson
&
Dennis M. Larsen

Washington State University Press
Pullman, Washington

Washington State University Press
PO Box 645910
Pullman, Washington 99164-5910
Phone: 800-354-7360
Fax: 509-335-8568
E-mail: wsupress@wsu.edu
Web site: wsupress.wsu.edu

First printing 2013

Library of Congress Cataloging-in-Publication Data

Johnson, Karen L. (Karen Leslie), 1955-
A Yankee on Puget Sound : pioneer dispatches of Edward Jay Allen, 1852-1855 / Karen L. Johnson and Dennis M. Larsen.
pages cm
Includes bibliographical references and index.
ISBN 978-0-87422-315-6 (alk. paper)
1. Allen, Edward Jay, 1830-1915—Correspondence. 2. Washington (Territory)—Description and travel. 3. Surveyors—Washington Territory—Correspondence. 4. United States. Army—Officers—Biography. 5. Military roads—Washington Territory. 6. Overland journeys to the Pacific. 7. Frontier and pioneer life—Washington Territory. 8. Cowlitz Trail (Wash.)—Description and travel. 9. Washington (Territory)—History—Sources. I. Larsen, Dennis M., 1946- II. Title. III. Title: Pioneer dispatches of Edward Jay Allen, 1852-1855.
F891.J57 2013
979.7--dc23

2013005507

Dedication

To my sister Serene, brother Garth, and best friend Karel Campbell for listening to endless snippets of Eddie's story. Thanks for your patience, love, and support. And to Vic Kucera, whose research request led to a vague reference to a young man on the Cowlitz Trail.

Karen L. Johnson

To my wife Pat, traveling partner on three cross-country treks in pursuit of Allen's story. She polished my contributions to this book well beyond my ability to do so.

Dennis M. Larsen

And obviously to Eddie, whose story, once scattered to the four winds, has come together at last.

Contents

Illustrations

Maps

Preface

Libraries can be dangerous. Libraries and the books they contain can take you to surprising and unforeseen places, consuming your life for long periods.

The spark that ignited our research into Edward Jay Allen's story was an article Karen Johnson came across while researching the Cowlitz Trail at the Washington State Library. That path ran from the Columbia River to Puget Sound, and was used by Native Americans, fur traders, soldiers, and settlers. Blanche Billings Mahlberg, daughter of a Washington pioneer, wrote an article in 1953 titled "Edward J. Allen, Pioneer and Roadbuilder," that excerpted several letters written by Allen during his trip over the Oregon Trail, up the Cowlitz Trail, and at his donation land claim in Olympia. The excerpts tantalized us. Mahlberg's excellent article spoke engagingly of Allen's travels, and his quoted writings came from the hand of an obviously talented and educated author, writing in the oh-so-descriptive language of the mid-1800s. We noticed a footnote stating Allen's letters had been published in the *Pittsburg Dispatch* between 1852 and 1855.

After a lengthy search, we located a scrapbook, held at the Beinecke Rare Book and Manuscript Library at Yale University, which contained clippings of Allen's letters from the *Dispatch*. Yale kindly provided us with digital copies of the scrapbook pages.

Over the next few months we transcribed the letters, which amounted to around 150,000 words. We became fascinated by Allen's adventures and roles in many of the pivotal moments in the early history of Washington Territory. Once we had finished the transcriptions, we started researching the people, places, and events that Allen described.

Our thought was to turn excerpts from the letters into a series of articles. We annotated one entire letter about the Cowlitz Trail and published the resulting article in the *Cowlitz Historical Quarterly*, a publication of the Cowlitz County (Washington) Historical Society. Meanwhile, we were researching and writing an article focusing on Allen's time spent surveying the route for the Naches Pass emigrant road over the Cascade Mountains. We had put together a draft of this article when serendipity intervened and we began thinking about a book instead.

In the spring of 2009, a series of history lectures was being organized in Olympia, and we were asked to present a talk on Allen. Luckily for all of us, the event was advertised on the web. Just two and a half weeks before the lecture, we received a phone call from a man in New Jersey. Gus Rosanio had found an online reference to the talk, and told us an amazing story.

In July 1981 Rosanio lived in a coastal resort town in New Jersey. On a walk to the beach one day he noticed preparations for an estate sale in progress at an old residence. A number of bags were placed at the curb next to the trash. Rosanio rescued two bags and took them home. Sorting through the miscellanea, he came across a large white envelope containing a bundle of letters dating from 1850 to 1854. The letters had been written on fine blue paper by various Allen family members to an Edward Jay Allen in Oregon and Washington territories.

Over the next twenty-odd years Rosanio transcribed much of the correspondence and became fascinated with the Allen family. He researched the histories of the various family members and even created a rough draft of a book based on letters written by Allen's sister Rebecca. Then in the spring of 2009 he decided on a whim to Google the name Edward Jay Allen. To his surprise he found an announcement of an upcoming lecture on Allen to be given in Olympia, Washington, by Karen Johnson. Full of curiosity, Gus picked up the phone and the rest, as they say, is history. After a century and a half, the two halves of Edward Jay Allen's correspondence were reunited.

As we began work on this book we were contacted by the husband of a descendant of Allen's who had been in charge of the estate sale preparations in New Jersey in 1981. Some young people he had hired to assist in cleaning the house prior to the sale, he told us, had inadvertently discarded the bags containing the Allen family letters. For the intervening three decades he thought the correspondence was irretrievably lost. Never in his wildest dreams did he imagine that someone had picked up the bags of letters, and that they would eventually become part of a book—this book.

We determined early on that one or both of us would visit as many of the locations mentioned in the Allen letters as possible. Without standing on the ground and looking at the geography with letters in hand, we felt our understanding of Allen's story would be incomplete.

The Northwest locations offered little in the way of hardship, as Dennis lives in Olympia and Karen just south in Chehalis, but they provided much adventure. We explored different sections of the Naches Pass Wagon Road. We ventured north to Whidbey Island to identify locations Allen cited while on his whaleboat voyage. Deciphering "Schroreckbin's Point" was a most interesting exercise. Karen had previously explored the Cowlitz Trail quite thoroughly when making a documentary film and knew that ground well.

Visiting the repositories of Allen-related letters, documents, and artifacts was another matter. We thoroughly searched all those in Washington State that contained relevant materials, but other Allen materials were scattered around the country in private holdings and at several universities. We contacted all of these sources by e-mail or phone, but felt that an in-person look at the collections would be best, as curators and custodians unfamiliar with Allen's story might have missed something pertinent despite our efforts to be thorough in our requests for documents. Accordingly, in late January of 2010, Dennis and his wife Pat undertook a 10,000-mile round trip to visit these locations across the country.

At a stop at the Hillman Library at the University of Pittsburgh, in a box titled "Allen Albums and Diaries," were several items that were no doubt the most prized of Allen's possessions, passed on to his descendants. Here were two miniatures of his mother and father, a small pocket compass, and a small New Testament inscribed as a gift from brother William to Eddie. Most likely the compass and Testament are the ones young Eddie carried over the Oregon Trail, and the compass is the same one he used to navigate over Naches Pass in 1853. Here also was Allen's 1852 Oregon Trail diary.

At the end of all this research, we concluded that we had enough material for two books: one on the Oregon Trail portion of Allen's journey, and another on his time in Washington Territory.

The Oregon Trail story is told in our book *Our Faces Are Westward: The 1852 Oregon Trail Journey of Edward Jay Allen* (Independence, MO: Oregon–California Trails Association, 2012). Allen's adventures in Washington Territory are told on the following pages.

Acknowledgments

The authors thank the following:

Gus Rosanio, for saving the Allen letters, finding us, and sharing his research and vast enthusiasm for all things Allen. We couldn't have done this without you.

Winthrop Baylies, and Elizabeth and Larry Cahill, who opened their own collections to us. They shared research and insights into their ancestor, and also shared many a family story.

The library and research staffs at Yale and Duke Universities for providing long-distance access to their collections, and the Universities of Miami, Pittsburgh, and Arkansas at Little Rock for hosting us in person.

Carol Garcia of New York for allowing us access to her grandmother Blanche Billings Mahlberg's research papers.

The staff at Washington State Library for keeping their wonderful institution (our "home away from home") open during troubled times.

Jewell Dunn, Charles Billings, Drew Crooks, and the late Roger Easton, for their welcome and invaluable assistance in research.

Historians, indexers, authors, and others who have added to our knowledge.

Edward Jay Allen, mid-1850s. Courtesy of Elizabeth and Larry Cahill.

Introduction

This is the tale of the adventures of a young man who in the years 1852 to 1855 journeyed across the continent via the Oregon Trail to the village of Olympia on Puget Sound, then part of the Oregon Territory. It is the story of the founding of Washington Territory and of that young man's part in the endeavor, of the building of a wagon road across the Cascade Mountains, and of life in a cabin on the shores of Puget Sound in 1853. It is the story of a whaleboat exploration of Puget Sound in the storms of December and much more.

Edward Jay Allen arrived in Portland, Oregon Territory, in the fall of 1852 after an arduous six-month journey from Pittsburgh, Pennsylvania.[1] After a brief stay he made his way north over the Cowlitz Trail to Olympia.

This twenty-two-year-old filled his two-year sojourn in the Northwest with accomplishments almost beyond description.

For the first time we learn the details of the 1853 scouting trip over Naches Pass that led to the creation and construction of a wagon road over the Cascades,[2] and that several "facts" historians have long believed about the Longmire wagon train, the first wagon train to use that road, are incorrect. We learn of the "leisure time" activities of the delegates to the Cowlitz Convention, which hastened the process of separating Washington from Oregon. Allen, writing in his enchanting style, introduced many of the who's who in early Washington history, such as Charles Terry, one of the founders of Seattle, and Isaac Stevens, the new territorial governor. Allen even participated in Washington's first territorial election. Perhaps the most amazing thing about Allen's tale is that it is mostly unknown to students of Northwest history.

We have been fortunate indeed to locate many unique primary documents around which we have built Allen's story.

1. Allen's 1852-55 Letters (The Yale Scrapbook)

The first set of documents to come to our attention was a series of letters written by Allen between 1852 and 1855. These letters were mailed to Allen's brother William, who excised personal and family information and sent the letters on to his hometown newspaper, the *Pittsburg Dispatch,*[3] which published them for an audience eager for news and information of the West.

Much to our delight, someone in the 1850s carefully collected most of the pertinent issues of the *Dispatch*, cut out the published letters, and pasted them into a scrapbook, now held by the Beinecke Rare Book and Manuscript Library at Yale University. This scrapbook seems to be the only extant copy of the vast majority of Allen's letters, in a public holding.

When Yale sent us digital scans of the scrapbook pages, we did not know who had created the book. Yale knew only that the Beinecke Library had acquired the scrapbook in 1966 from a dealer in Western Americana, and that the dealer was from Fort Smith, Arkansas. As our research proceeded, we wondered who had carefully preserved all those articles. No clues were found, though, until Yale posted the scanned images online, and titled the collection "Oregon Trail / Eddie's Letters." An email to Yale solicited the information that the phrase "Eddie's Letters" was handwritten inside the scrapbook. Who would use that phrase except a relative or close friend?

We laid that question aside for a time as we researched other issues, one being items related to Allen's sister, Rebecca, whose married name was Turner. We found Turner documents at the University of Arkansas, Little Rock, and at Duke University in North Carolina. We asked Duke to copy some relevant correspondence from their Turner Family Papers collection. This material included a letter written by Allen's brother, William, in early December 1854.

At the time of that writing, Edward was living in Olympia, Washington Territory, sister Rebecca was living at the family home in Pittsburgh, and brother William was working on a steamboat on the Mississippi, Ohio, and Arkansas Rivers. In the letter dated December 4, 1854, William wrote to Rebecca about the upcoming Christmas holiday. He proposed gift ideas for various relatives, and finally asked, "And now I am at a loss for you Rebecca, what shall I get? a card case, or a scrapbook for Eddie's letters?" Well, what a find! Could it be that William did indeed give Rebecca an empty scrapbook for Christmas, a scrapbook in which she could paste Eddie's letters that had been published in the *Dispatch*?

A re-read of the acquisition information from Yale verified that the dealer who sold the scrapbook was from Fort Smith, Arkansas. When Rebecca married Jesse Turner, she moved to Van Buren, Arkansas, just a few miles from Fort Smith, and lived there until her death in 1917. Although we may never be able to *conclusively* state that Rebecca created the scrapbook, we can surmise that she did, and that after her death, the scrapbook passed to a relative or friend and eventually into the hands of the document dealer.

"Eddie's Letters" became the heart and soul of this work. We refer to these letters as the Yale Scrapbook in our citations.

2. Allen's Pittsburgh Manuscript

Sometime in the early 1900s, Allen decided to turn the original letters published in the *Dispatch* into a book he titled "Over the Oregon Trail," though it appears this book was never published. His typewritten manuscript[4] included revisions and clarifications that we found very helpful in solving riddles of spelling and deciphering names of people and geographical locations. The manuscript most often enhanced the story with additional details and new information, but at times it added a note of confusion when Allen contradicted himself or seemed to remember incorrectly. This document, like the letters, was incomplete: the manuscript ended before Allen's whaleboat voyage on Puget Sound in December 1853.

Allen's manuscript also contained a typewritten dictionary, compiled by Allen himself, of the Chinook jargon, a trade language used in the Pacific Northwest and composed mainly of French, English, and Native American words. In our notes, we give the English translation of jargon words, but have foregone more extensive citations, as all translations are taken from Allen's dictionary or that of James G. Swan in his book, *The Northwest Coast* (see Bibliography).

While the Pittsburgh Manuscript's introduction is dated October 1913, we know from

Allen's correspondence with Ezra Meeker that it was actually written several years earlier, most likely around 1908.[5]

3. Allen Family Letters and Materials

A large number of family letters written to Edward during the time when he was out West, and a few written by Allen himself, are held in the private collection of Gustave Rosanio of Riverside, New Jersey. Rosanio has kindly granted us unfettered access to his collection. We refer to these letters as the Rosanio Collection in our citations. Another set of family letters is held at the Rare Book, Manuscript, and Special Collections Library, Duke University, under the title Turner Collection; we refer to these letters as Turner Collection, Duke, in our citations. A similar collection of letters may be found at the Ottenheimer Library, University of Arkansas Little Rock, also under the title Turner Collection; we refer to these letters as Turner Collection, UALR. Finally, Allen descendants Winthrop Baylies and Elizabeth Cahill have shared a few related documents.

A Note on Methods

In the preparation of this manuscript, we used the Yale University scrapbook as our primary source. We knew that the letters in the scrapbook had gone through at least two transcriptions before appearing in the *Pittsburg Dispatch* in the 1850s. These letters were written to Allen's brother William, who then excised family and personal matters from the text and sent the edited letters to the *Dispatch*. The originals were forwarded to Allen's parents and other siblings. The editor of the *Dispatch* turned his copies over to a typesetter who no doubt made a number of errors when confronted with strange Indian and local place names, and other personal names. Nevertheless, as these were the accounts of events closest in time to their actual occurrence, errors in transcription notwithstanding, we decided to base our story on them.

Conventions of writing, spelling, and punctuation have changed during the century and a half since the letters were written. For example, many current words spelled with the letter "z" were then spelled with an "s." In another example, the term "etc." was not in common usage in the 1850s. In its place "&c." was used to convey the same meaning. We have left words and phrases as they were originally written as long as they are easily understandable. We use brackets [ ] to correct spelling and typesetting errors only when a correction or clarification is needed to understand the intent of the writer. We have kept the use of these to a minimum.

We have incorporated small parts of Allen's Pittsburgh Manuscript into our story. When this manuscript cleared up a single word, we continue to use brackets to clarify, not setting it off in any special manner. However, when we use longer portions of the Pittsburgh Manuscript in the text we identify it as such or place the excerpts in braces { }.

A Note on Poetry

Considered one of the greatest essayists in the English language, William Hazlitt (1778-1830) pioneered the style of inserting quotations of various kinds, often poetry, into his writings. The use of this technique in letter writing became widespread in the United States in the 1840s and early 1850s. Allen, a la Hazlitt, sprinkled his letters with poetry and in the process demonstrated the depth of his classical education. Shakespeare, Lord Byron, Longfellow, and unknown authors published in *Knickerbocker* magazine[6] all made their way into his letters. Allen quoted the poems from memory. He confirmed this in a letter written just before he arrived in Portland: "It seems to me, by the by, that crossing the plains has quickened my memory, and enabled me to remember everything in the poetical line I ever read, a fortunate thing too, as one don't get the originals out here."[7] Poetry was Allen's life-long love. When he returned to Pittsburgh he wrote lyrics for songs and eventually published a book of poems, some of which referred to his time in the West.[8] We have presented the poems and quotes in Allen's letters and Pittsburgh Manuscript with his original punctuation and wording. Since the poems were quoted from memory and often diverge from published versions, we have not provided specific citations, but have identified the poem and author when possible.

Racism in the 1850s

In his descriptions of the native peoples he encountered, Allen sometimes appeared to be demeaning or casually patronizing, using stereotyped language and references, but seemingly without malice. Not having much or any contact with Indians at home, he was newly exposed to and probably influenced by prevailing attitudes (and vocabulary) toward native peoples that could at times be intolerant, racist and derogatory, and often overtly hostile. This was not Allen's view. His letters indicate his clear understanding that westward expansion, with its attendant diseases, was destroying Native American society, and he regretted it. He respected the Indians with whom he interacted, observed their culture, and made an impressive study of the Chinook jargon. He believed, contrary to prevailing sentiment, that the tribes had rights of ownership and should be compensated for the loss of lands they had inhabited for generations.

Allen's Way West

"Ah! this Oregon is a beautiful country! I know of no place where a man could live with more pleasure or with less care for the morrow than here." [1]

The Allen Family

Edward Jay Allen's road to Puget Sound in far-off northern Oregon Territory actually began in 1827 on the other side of the Atlantic Ocean in Warwickshire, England. That year his father, Edward Allen (b. 1794), and his mother, Amelia (Millicent) Bindley Allen (b. 1797), immigrated to the United States with their four children: Elizabeth (b. 1817), William (b. 1820), Rebecca (b. 1823), and George (b. 1825). The family settled in New York State, where two more children were born: Amelia in 1828; and the hero of this story, Edward Jay Allen, on April 27, 1830.

By 1840 the family had moved to Pittsburgh, Pennsylvania. Family correspondence depicted the six Allen siblings as well educated, articulate, and very close-knit. It also revealed that the elder Allen operated a construction business involved in such diverse projects as building a hotel, creating a sewer system for the city of Pittsburgh, and laying the brickwork in the summit tunnel for the railroad then making its way over Pennsylvania's Allegheny Mountains.

Edward Jay Allen's older brothers William and George had engaged in steamboating for some years, learning the trade and eventually forming a partnership and purchasing the steamer *New Hampshire*. The brothers operated their boat on the Mississippi River and its various tributaries. William was captain, and George acted as the first clerk. On the sixth of May, 1847, about forty miles below Little Rock, Arkansas, the boat caught on a submerged snag, and during an attempt to free the steamer, the boiler exploded. George was killed along with many others; William was stunned. This tragedy colored the Allen family correspondence for years to come. After recuperating, William returned to steamboating and young Edward occasionally joined him. During the period of Allen's westward adventures Elizabeth was the only married sibling, living in rural Fairfield, Pennsylvania, some fifty miles east of Pittsburgh. Rebecca and Amelia were living in their parents' home and teaching in the Pittsburgh public schools.

Edward Allen attended Pittsburgh public schools and Duquesne College (an offshoot of the Western University of Pennsylvania and a forerunner of the University of Pittsburgh), where he studied the classics, including a thorough grounding in poetry.

Allen worked in a grocery warehouse during his school years.[2] On his days off he occasionally took a steamboat up the Conemaugh River to visit his sister Elizabeth and her husband Samuel Fundenburg. Here he dabbled in the study of medicine under the tutelage of his brother-in-law, a country doctor. This study came in handy along the Oregon Trail and during his time in Washington Territory.

Allen was well acquainted with doctors. For some time he had been in ill health with a severe cough that caused much spitting up of blood. He wrote that the ailment "had almost entirely closed my windpipe—so that after a few sentences I would assume a voice like a quavering trombone." The cough, "like one of the big Brass affairs in a Dutch Band," left him "worn out weary."[3] While traveling with his brother William on the steamboat *Childe Harold* Allen wrote a letter to his family in Pittsburgh in which he described his chronic illness.

> I feel almost relieved of my cough. I got a prescription filled in Cincinnati which my friend the Dr gave me—and it has done me more good than anything I have yet used. My throat is not in the least sore and my cough appears to have become loosened. I hope soon to drop the possessive and [no] longer speak of "my" ailments. I want no more of them—if they will not disappear from the earth altogether let somebody else have them—I have had my day, and a thundering long one it has been.[4]

Allen's ailments were still "his," however. On March 27, 1851, his sister Amelia wrote, "[Dr. Wilson] also examined Eddie with a what you may call it and pronounced his lungs sound, not at all predisposed to consumption. He told him he had a splendid chest and must take great care of himself. This Ed told me but I think there was something said, judging from Edwards manner. The spitting of blood has entirely ceased, but when he reached New Orleans, he says he knows he lost as much as a pint of blood. He was very much alarmed at first."[5]

Allen spent the summer of 1851 in the hills of Westmoreland County, Pennsylvania, where the fresh air and outdoor life somewhat improved his health and spawned the idea for a journey west.

> At the age of twenty-two I found myself, if the doctors' opinions were to be accepted, very dilapidated as to lungs and in a general condition of goneness that did discredit to several generations of sturdy ancestors. An outdoor summer in the hills of Westmoreland County, and its resultant benefits, gave the cue for a more extended outing, from which grew the idea of a trip to Oregon. It seemed a kill-or-cure remedy and offered adventure. Its promise of danger was not to be seriously considered by a man who carried in his own breast a danger that admitted of no doubt.
>
> A study of all the available information upon the necessities of such travel resulted in the fitting out of a wagon in Pittsburgh in association with three companions,[6] shipment of the outfit by river to St. Louis, the purchase of some yoke of cattle and complete supplies there, shipment by steamboat to Council Bluffs, purchase of additional oxen, crossing the Missouri River to what is now Omaha, and so across the plains to Portland, Oregon, some nineteen hundred miles over "The Great American Desert."
>
> The journey was faithfully chronicled in letters home which were, with the omission of personal matter, published serially in a local newspaper during the period of three years from 1852 to 1855.[7]

Thus in late March 1852, a somewhat sickly twenty-two-year-old Edward Jay Allen began a venturesome journey across the continent to the village of Olympia on the southern reaches of Puget Sound. Allen's decision to go to Puget Sound may have been influenced by correspondence with Quincy Brooks, also of Duquesne and Pittsburgh, who had already made the trek to the Northwest.[8] Allen's "kill-or-cure" solution fortunately resulted in a cure.

The Oregon Trail

The four Pittsburgh friends made their way to St. Louis, arriving at the end of April. Here they transferred their wagon, supplies, and newly purchased oxen onto the steamer *St. Paul.* They departed St. Louis around May 1 and arrived in Council Bluffs on May 11,[9] where the "city boys" encountered great difficulty in unloading their oxen. One unruly beast dragged them all over the prairie before finally being subdued. Camp was set up near the steamer landing four miles from "town," where they had to carry their wagon tongue to be strengthened. A sick and weary Allen crawled into his covered wagon bed that night and spent the wee hours looking at the stars and thinking of home.

Four days after arriving, the Pittsburgh boys drove their wagon to the ferry and obtained permission to join a train of thirty-two wagons made up primarily of farm families from Iowa. They hoped to cross the Missouri River in the morning.[10] It was not to be. A high wind made the river too rough to cross, so they drove the cattle three miles back to grass and waited. During the wait Allen and Jacob Resser went into town to purchase a third yoke of oxen. While there they had their pictures taken at a daguerreotype studio and mailed the images home.

Cooking offered an additional challenge to the young men. One of the four (Allen did not say which one) volunteered to cook, and they agreed that the cook would be relieved of guard duty and oxen care. However, after a sampling of the fare, Allen concluded, "Maybe when we have been a month or so on the trail and his ability is measured by comparison with other 'trailers', he may rank high; at present, where he is judged by the standard of home cuisine, we are not considering the presentation of a medal."[11]

Camp life was new to the Pittsburgh boys and they took twice as long to do chores as did the seasoned farmers. Each night they dropped like the dead into their wagon to sleep. After a week they were finally ferried across the Missouri River.

The company traveled on the north side of the Platte River following what is known today as the Mormon Trail. Allen wrote of prairie storms and stampeding oxen, exorbitant ferry fees at river crossings, the Pawnee Indians stealing horses, cooking over buffalo chips, and the everyday hardships of the trail.

Allen wrote his letters home as opportunities arose—from his wagon whenever there were pauses in the journey, at the evening campfire, and, at times, sitting on his horse. The letters grew to many pages and he mailed them wherever he could—from forts along the way (Laramie, Hall, and Boise) or with Blodget's Express at such remote places as Crab Creek, Nebraska, and Devil's Gate, Wyoming.[12]

Twenty-six days and 522 miles into the journey Allen's company arrived at Fort Laramie, Wyoming. As an ardent reader of *Knickerbocker* magazine, Allen was eager to see the places where the publication's western heroes roamed. Fort Laramie was high on his list. He said of the fort, "It is a very pretty little place... The public buildings are a blacksmith and wagon shop, a bakery, a store, and the Post Office."[13] Allen took the company's letters across the Platte River via ferry, mailed them at the post office and bought some supplies—fewer than he intended, as he was appalled at the store's high prices. Just west of Fort Laramie Allen severely sprained his ankle; regardless, riding in the wagon was not an option. Contrary to the Hollywood version of the Oregon Trail, only infants and the seriously ill or injured rode as passengers. A "driver" walked beside the oxen, controlling them with a goad. Supplies filled the wagons and loads were kept as light as possible to avoid stressing the oxen. At Devil's Gate Allen spent forty dollars on a riding pony, but found he often had to lend it to the scouts who preceded the wagon train looking for good grass, water and suitable campsites. To make matters worse, the company had been racing for weeks to get to the head of the emigration. Those at the rear found the grass already eaten when they stopped to camp at night, which required them to drive their stock several miles up side canyons to find suitable feed. Allen's company decided to sprint to the front rather than spend every night and morning doing this, and the ambitious pace was brutal on his ankle.[14]

On June 24 the company reached Independence Rock, Wyoming. Allen climbed to the top and took in the expansive view. On July 1, at noon, he reached South Pass (Wyoming) and noted this milestone that marked the crossing out of the United States and into Oregon.[15] At the "Parting of the Ways" the company took the Sublette cutoff rather than the gentler but longer route via Fort Bridger.[16] On July 6 they reached the Green River at Names Hill, where several party members left their inscribed graffiti,[17] as did thousands of pioneers on rocks and boulders along the entire route. From here Allen began the long climb up Dempsey Ridge, at over 8,000 feet the high point on the trail. At the summit they encountered a true scourge.

> You would hardly suppose we would be afflicted with those pests of the summer, musquitos, here—neither did I, and in my omission of this, one of the hardships of the route, I omitted the greatest. I have "suffered some" from their voracity, at home—and more in their native haunts, along the banks of the sullen Mississippi—but the musquitos here, in bloodthirsty and dogged perseverance, exceed all hitherto known.[18]

Allen's company continued on to the Bear River, Soda Springs and Fort Hall (Idaho), then followed the Snake River (or the Lewis River as it was then known) in the heat of August. One morning Allen's pony wandered off and, despite a vigorous search, was not found. Allen's bad ankle now presented a serious hazard to his survival.

At the ford of the Snake River the company found no ferry and so they started down the south bank of the river, taking what is today known as the south alternate trail. Allen noted a couple of parties caulking their wagon beds and making them into boats, obviously with the intention of floating down the remainder of the Snake River. Despite the heat, the dust, and the misery of August in southern Idaho, few pioneers chose the river option—with good cause. The next day, however, Allen decided to cast his lot with the river.[19] With his ankle injury, walking was now out of the question, and riding in the wagon would wear out his oxen and thereby jeopardize the safety of the rest of the company who would be obliged to care for him. It was the river or nothing. He gave Moses Hale of Iowa the use of his wagon's running gear and oxen and began building a boat out of his wagon bed. Several others joined him and two more wagon-boats were built. On August 8 Allen and eight other men, one woman, and two children started down the river.[20]

Allen's recounting of this unique and dangerous journey can be found in *Our Faces Are Westward.*

On August 18, ten days and some 200 miles after launching, the wagon-boats landed safely at Fort Boise. Everyone had had enough of river travel. Allen and some of his fellow floaters settled in for a needed rest and found a good way to put their wagon-boats to work—they founded the Pennsylvania Ferry Company. Until September 12, when Allen shut the business down and continued his journey west, they earned good money ferrying emigrants across the river. A few days before he quit ferrying, Allen noted that two of his partners in the ferry business by the name of Meeker had sold out their interest.

Ezra and Oliver Meeker were also journeying west from Wapello County, Iowa, and Allen joined their ferry enterprise at Fort Boise. Meeker later described the ferry business:

> My friend had crossed the plains the same year I did, and although a single man and young at that, had kept a diary all the way. Poring over this venerable manuscript one day while I was with him, Mr. Allen ran across this sentence, 'The Meeker brothers sold out their interest in the ferry today for $185.00, and left for Portland.' Both had forgotten the partnership though each remembered their experience of ferrying in wagon-boxes.[21]

Meeker was in error. Allen's August 21 diary entry clearly reads, "Good fellow Oliver Meeker… Meeker sold out for $85." Meeker must have extracted his not quite accurate quote from Allen's 1852 letters, not from his diary. The incorrect amount of $185 was printed in the *Dispatch*. Allen stayed at Fort Boise a week or two longer, continuing the operation of the Pennsylvania Ferry.

Allen used some of the proceeds from his ferry to buy a horse and a mule. On September 12, in company with Monroe Baldwin (a partner in the ferry business) who was also on horseback, Allen started on the last phase of his journey west. Waves of emigrant wagon trains had cleaned out the provisions at Fort Boise, and the two men secured only three days' worth of food before they departed. They hoped to purchase more food from the wagon trains they passed along the trail, but this hope turned out to be foolishness, as by this point the emigrants were all at the end of their rations and had none to spare.

Near today's Powder River, Oregon, Allen's exhausted horse faltered. He gave the worn-out horse to a "Californian" in a passing wagon train in hopes that by traveling at a slower pace it could be nursed back to health. Allen then switched to his mule. By this time Allen and Baldwin were living on what little they could shoot.

The two hungry men worked their way up to the crest of the Blue Mountains and down the west slope to the Umatilla River. Here they traded Baldwin's worn-out horse to a packer for a nearly wild pony and spent a good part of the day breaking the animal. Shortly after, Allen traded his worn-out mule for a fresh Indian horse. Baldwin put a saddle on his pony, mounted, and was immediately thrown. Pony and saddle galloped off, never to be seen again. That night they tied Allen's horse, their only remaining animal, to a large boulder with a lariat so that he could wander a bit and graze. In the morning they found the lariat still wrapped around the boulder but the horse was gone. They were, as Allen wrote, "300 miles from anywhere and horseless."

Putting their remaining possessions into their saddle packs, the nearly starving duo walked twenty-two miles that day. By the time they reached Well Spring, in the middle of the eastern Oregon desert, they were nearly done in. A sympathetic but destitute emigrant offered them a hambone all but denuded of meat. That and a bit of water from the spring moved them on a little farther. Six miles from the next water, Baldwin collapsed; Allen, barely able to stand, flagged down a passing wagon. Aware that his companion was an Oddfellow and a Mason, Allen inquired if there were any such in the wagon train. A member of one of the fraternal orders came forth, loaded Baldwin into his wagon, and proceeded on. Allen was left at the roadside.

The Trip to Portland

Allen's hard cash, earned working the Pennsylvania Ferry, bought him a series of horse rides for the remainder of the way to The Dalles. His survival was no doubt aided by his youth.

Upon reaching The Dalles, Allen wrote simply, "All that was left was a boat trip down the Columbia." Yet true to the dramatic nature of Allen's trek, this journey of eighty miles offered its own excitement. Allen's October 15 letter, excerpted here, was written after his arrival in Portland, and described the last few days of the journey west.

> That evening we came to the "Dalles"[22]—the long wished for termination of our journey by land, for I was determined not to cross the Cascade Mountains so late in the season, and water near. There were collected here more than 500 wagons, the owners of which were apparently of the same opinion, and were waiting to go down the river in boats. It was a long sandy beach, and the sand was blowing a perfect simoon;[23] we toiled on down toward the camp, which ever and anon we could catch a glimpse of in the hill of wind.[24] I at last had to make a dead halt, our horses (or rather mules) wheeling to face the weather; my eyes were literally bunged up with dust, and (as I sat blinded with it my head bowed down over my saddle, waiting a cessation of the wind,) suddenly I heard a Yankee-like nasal voice cry out, right under me, "Want any pies? Pies!" My eyes opened of themselves—wide, in sheer astonishment. The wind had changed, and sure enough there, clear and distinct before me, stood a speculative-looking individual, with a tin tube-like box under each arm—who had addressed "Pies!" to a man who thought a herring was a dissipation. I was into the contents of that tin pannier quickly, and didn't leave him till he pointed out his shop. Although nearly two feet in circumference, there wasn't much material in them; as honest old John Browdie said of the diet at Detheberg school—"they only aggravated a man, instead o' satisfying 'im."[25]

East of The Dalles, Allen encountered a nearly destitute couple (Isaac Smith and his wife[26]) in the midst of lightening their load of remaining possessions. Smith's wife fussed greatly at having to abandon her skillet and many other items by the roadside. Allen came to their aid, purchased some food for them and accompanied them to The Dalles, where he arranged their passage down the Columbia River: a two-day float to the Columbia River Cascades[27] in an open scow, buffeted by strong winds and rain, to a landing on the north bank of the river. Here Allen encountered several members of the Iowa wagon train waiting for the steamboat *Multnomah* to carry them the final miles to Portland. Allen was quite surprised when they returned his oxen to his care. The original bargain was that the train would get the oxen to Portland in exchange for their use from Three Island Crossing. Intending to also sail on the *Multnomah*, Allen hired a young boy to drive the oxen the final miles west on the pack trail from the Cascades. His intention was to take possession of the oxen in Portland. Events determined otherwise and nearly a month later he found them in the vicinity of Fort Vancouver.[28] On Monday, October 4, Allen, Monroe Baldwin and many from the Iowa wagon train boarded the *Multnomah* and steamed downriver to Portland. The Smiths, with tickets purchased by Allen, came on the next sailing a few days later. Allen's October letter continues:

> Passed Fort Vancouver, where a great many emigrants were remaining over winter. The Columbia, below the cascades, is a most magnificent stream, two or three miles wide, with beautiful scenery—placid and peaceful, and some of the tall trees you "read about." We came to the mouth of Willamette River,[29] a small stream, not as large as the Monongahela,[30] passed up, leaving the Columbia, a ship was going up also, with the tide, and several were at the landing, and a respectable looking steamer, the "Lot Whitcomb."[31] I met many acquaintances on the boat—De Ballard[32]

and lady, besides others, and more in the city, where we landed just at dusk; all went to the "Yarn [Yam] Hill House" together, and I felt like drawing a long breath, that I had at last *got through*. Ate a tremendous supper, more than a christian man ought to eat, and felt like lying down and sleeping six months; *did* lay down, and in my dreams retraced all that five months weary way over again, threading back, day by day, each days progress—till I was again at home with you all—which was my first waking thought on my first day in Oregon, succeeded by the involuntary mental inquiry of when this dream would be realized? No longer a time, I fervently believed, than the period set when I bid all good-bye.

Fort Vancouver, Washington Territory, by Gustav Sohon. Library of Congress Prints and Photographs Division, Item #2011647869.

Portland and the Cowlitz Trail

Allen had arrived, more or less intact, in Portland. His stay here would be brief as his goal was to reach Puget Sound. A trip over the Cowlitz Trail, driving his oxen, would be interrupted by the unexpected duty of attending the Monticello Convention, a gathering that hastened the process of separating Washington Territory from Oregon.

The following letter tells the story of Allen's arrival in Portland and travels north.[1]

Portland

Thus far have I come, through "the bowells of the land," unto my journey's end, even unto a "City of Habitation," having, since the date and locality of my last letter, (September 13th) Fort Boise, come some 500 miles of the most wearisome portion of the whole journey, with less energy and strength to endure its hardships.

It is a pretty little place—the largest city in the Territory, containing perhaps about 1,200 inhabitants, pleasantly situated on the West bank of the [Willamette] river, the adjacent streets, running parallel to which have side walks and crossings of plank, which are easily made out of the straight pine growing here, which will split in interminable lengths and of any degree of width as "straight as a shingle." The "respectable" looking blocks of stores and dwellings—with three or four Churches interspersed—looked very grateful to my eyes, as (exchanging the lonely prairies and black mountains,) it emerged into the appearance of something like civilization and human comfort.[2]

Nature of the Settlers

After a few months, nearly, of free, unrestrained mountain life, the transition to the civilized appearances of things around me was very abrupt and strange. I felt "cabined, cribbed, confined," in the garments custom made it obligatory on me to wear—and I might say propriety, too, at sight of my discarded coverings—and ill at ease, awkward and constrained, with the conventionalities the wearing of them imposed on me. I haven't "got the hang of them yet," or recovered my old ease of manner, and found it necessary to be perpetually apologizing to people I "bump" in the street, as I roll through, in the rollicking, arm-swinging way one unconsciously acquires on the Route. The apologies are received kindly, but with a latent smile, by the frank-hearted people one impinges against, who look you square in the face as men ought, and with the air of *men* who have breathed the atmosphere of "free soil—free speech," and who are emphatically "free men."

They *are* a great people—no finer colony exists on earth *anywhere!* They are the cream of New England, the "salt of the earth," men of mark, universally; and have set an example, in their settling and colonizing this part of the earth, that is unexampled in history. The first thing they did, in their legislative capacity, was to pass laws doing justice to the Indian (in that exceeding the example of our Quaker-hanging and Baptist-banishing Puritan ancestors,) then they passed a "Maine Law,"[3] and during the time I have been here I have not seen a drunkard on the street. At the same session they made a law prohibiting any man from owning more than 640 acres of land—ample space enough, God knows, when a man with ten acres won't thank you for any more—and crowned all by the passage of an act *prohibiting forever the introduction of Slavery within their limits, in any shape!*[4]—an act which should ensure to them the love and respect of every good man in the land. And now, with *such* a people, industrious, thinking, God-fearing men—removed leagues from the contaminating influence of an effete civilization—their "lives cast in pleasant places," in a land "flowing with milk and honey," where the seed scattered broad cast on the scarcely turned soil is ripened by a sun so genial and atmosphere so ethereal that it rivals the climate of Italy or the fabled airs of "Araby the blest"—with a wide extent of country, undulating away to the snow-capped Rocky Mountains—where cattle require no shelter all the mild year round—with gentle running streams, flowing over pebbly bottoms, irrigating a great part of the country—bays on the coast wherein a "seventy-four"[5] can ride miles into the interior, and safely lay at anchor, with her masts and bowsprit piercing the foliage of the immense forest, casting a shade over the clear waters. What shall hinder this land from rivaling in power and riches, and equal distribution of comfort, that of the Jews in their palmiest days, when a territory of only one by two hundred miles in extent supported a population of incredible millions?[6]

A Problem with Dates

So young Allen finally reached the Pacific coast. But his transcontinental trek was not over. His next letter, dated December 22, 1852, was written after he had "progressed one hundred and fifty miles further" from Portland to Olympia.

The date of this letter presents an interesting problem. First, some background on the formation of Washington Territory is needed.

In 1852, Oregon Territory encompassed both present-day states of Washington and Oregon. The mighty Columbia River, running from east to west, served as a natural dividing line between the two halves of the territory, which were referred to as Northern Oregon and Southern Oregon. Settlers north of the Columbia River had been agitating for a separate territory for some time, feeling that their interests were poorly represented by the territorial government headquartered in Oregon City, near Portland, and later moved even farther away to Salem. Accordingly, in August of 1851, delegates from the Puget Sound area met at Warbassport/Cowlitz Landing to coalesce their thoughts and send a formal appeal or "memorial" to Congress, requesting a "legal divorce" from Southern Oregon, a request that Congress largely ignored.

Undeterred, the citizens of Northern Oregon organized the Monticello Convention of November 25, 1852, to write a memorial with more logic and less vinegar. Representatives from Southern Oregon, however, were already on the verge of asking Congress for the separation. Southern Oregon residents wanted to achieve statehood, and the combined northern and southern portions were considered too large to be admitted as one state. Territorial Representative Joseph Lane introduced the separation bill in December 1852.

Allen was asked to attend the Monticello Convention, and did so willingly and proudly. The meeting resulted in a second memorial to Congress, again requesting the creation of a new territory. The delegates even suggested a name: the Territory of Columbia. The memorial was duly sent off to Washington City (Washington, DC) through regular channels, which in those days meant a trip by ship from the Northwest down the west coast of America, by land across Nicaragua or the Isthmus of Panama, then another voyage across the Caribbean and up the eastern seaboard. Average delivery time from the hinterlands of the Northwest ran several weeks.

Lane's bill passed the House on February 8, 1853. The Senate approved the bill on March 2, 1853, and immediately thereafter President Millard Fillmore signed the act creating the new territory. The momentous news first reached Puget Sound in the form of a rumor.

On April 9, 1853, Olympia's newspaper, the *Columbian* published a letter from Oregon Territory Representative Joseph Lane:

House of Representatives, Washington City, D.C., January 31, 1853. Quincy A. Brooks, Esq.: —Dear Sir:—I have the honor to acknowledge the receipt of your favor of the 3d December last,

enclosing a Memorial of the Delegates of the citizens of Northern Oregon, praying a division of the Territory of Oregon, and the establishment of the Territory of Columbia. Some time since, I submitted to the consideration of the House of Representatives, a Resolution proposing the object sought by the terms of your Memorial. The Resolution was referred to the House Committee, on Territories, who, on the 25th ultimo, reported a Bill I had drafted for the purpose, a copy of which I herewith enclose you. You may remain assured of my efforts for the passage of the Bill at this session, though I cannot promise that my labor in this respect will prove successful. Your obedient serv't,
JOSEPH LANE

In the same issue of the newspaper, the editors added this paragraph: "LATER.—Since the above was put in type, we have seen a gentleman from Portland, who states that the steamer had arrived, and a rumor was current there that the Territory had been divided, and the name 'WASHINGTON' had been substituted in place of "Columbia."[7]

Thus, on April 9, 1853, the residents of Oregon Territory learned for the first time, and then only by rumor, that a division of the territory had been approved, and that the new entity north of the Columbia River would be called Washington.

Formal word reached the Northwest on April 25, 1853, prompting a spontaneous and heartfelt hundred-gun salute in Olympia. Washington Territory was a fact.

However, Allen's letter was dated December 22, 1852. In this letter he wrote of the successful division of the territory, and that Washington had been chosen as the name. But Allen could not have known this in December 1852, as it had not yet occurred.

Also in this letter, Allen described Northwest wildflowers—trilliums, lupines, the fragrant honeysuckles and roses. His letter began with events of October, when he first arrived on the coast, and ended in late December. But during this time period he would not have seen any spring-blooming wildflowers.

How to explain this confusion? In a letter dated June 24, 1853, brother William wrote to Rebecca explaining his arduous task of editing Edward's letters.

At Cincinnati I secured Eddie's long expected letter containing the missing sheets, and so anxious was I to have it published, I worked at it while in port as I wrote you, and finished it [its] 20 pages and mailed from Louisville. Then commenced bout writing again for the present trip and finishing it, dashed at the "Continuation of the Oregon" Trail from Portland to Puget Sound...On my up trip I received the "Oregon Trail extended to Puget Sound" and finished it accompanying 24 more of these pages and have it ready to mail when we get up, if we do get up, for the water is very low...Now, when you recollect that in addition to the mechanical labor of writing, is to be added the task of omitting some parts, substituting others and transposing many to make them read connectedly you will have an idea of how much I have done.

By combining and editing two or more letters, and sending the revisions on to the *Dispatch* for publication, William was likely responsible for the confusion of dates. No matter—Allen still wrote a wonderful description of a ghastly trip up the Cowlitz Trail.

Olympia, December 22, 1852.

My last letter, written at chance times underway, partly at Oregon City and Portland, and finally mailed at Milwaukie, brought me to the settlements, "even unto a City of Habitations," and concluded the series of letters from the "Great Plains."

You will see from the heading of this letter that since that time I have progressed one hundred and fifty miles further, through the bowels of the land, which journey, when is taken into consideration the season in which it was undertaken, the length of time consumed, and the exposure and hardship it involved, may very properly be considered an extension of the "Oregon Trail."

For a week or two after my arrival in Portland I did little else than take "mine ease in mine own Inn," my time being occupied principally in bathing my ancle, making enquiries for a medical dictionary, to see what the meaning of Bronchitis was, and trying to restrain within decent limit my appetite, which in voraciousness had become absolutely unnatural. I embraced the opportunity of renewing old "Trail" and "Ferry" acquaintances and forming new ones. The letters that friends had been kind enough to give me, so that "in strange eyes I might not be a stranger," were of great service, bringing me in contact with persons whose advice was important, and friendship desirable. Governor Gaines[8] (to whom I had many letters, from Gen. Lane,[9] Thos. Corwin,[10] and Col. Black,[11] among others,) received me with cordiality. He spoke of the pleasure the letters afforded him, in being from personal friends, whom he had not seen for several years. He also evinced a flattering interest in my welfare, and gave me much useful information and advice. His view of the great advantages possessed by Northern Oregon over the Southern portion, and especially of the high destiny of Puget Sound, confirmed the impression I had previous to leaving home, and deepened the resolve I had formed to go on, as soon as my cattle should arrive. Our interview was very friendly and pleasant; he asked me frankly what I intended to do. I answered as frankly that I had come for health, and having obtained it, I felt inclined to do whatsoever my hands found to do, and was not at all choice in the selection. His answer was, emphatically, "I would do for Oregon." Oregon shall *do* for me.

As soon as I had recovered the use of my ankle, I spent my time in rambles round the country, sometimes by steamer to different points, oftener by canoe, and oftenest on "Shank's mare,"[12] and by this means was enabled to see a great deal I otherwise would not have seen, and feast my eyes with beautiful landscapes and delightful scenery. It would be very apposite here to give a concise account of the country, as I have seen it—its advantages as an agricultural and grazing country, the fertility of its soil and salubrity of its climate, but this would involve more time in the elaboration than would suit the free gossipping style of my letters to you, written on any and every occasion—never

"—knowing what next I'll say,
Writing what's uppermost without delay."[13]

I have wished some time to write such a letter, for publication, as, even written in my stuttering way, (embodying, as it would, the results of personal observation,) would be of service to the many proposing to make Oregon their future home; but want of time has hitherto prevented me. I only wait, however, the "mellowing of occasion" to put my resolution into effect, and, meanwhile, my letters will, I hope, answer the purpose for which they were written, viz: to give you an idea of my whereabout and doings.

A Logging Enterprise

Allen's contemporary December 22, 1852 letter to his brother William Allen continues with sections of the Pittsburgh Manuscript interspersed and identified with braces { }.

Feeling, with my increasing strength, a desire to engage in something till my cattle should arrive, and chancing to come across the proprietors of the Portland saw mills, they were complaining of the trouble of getting a supply of saw logs. I proposed, at once, to supply them, and, incredible as it may seem to you, with your ideas of my physical proportions, I set to work, and soon had a raft cut and delivered in Portland. Poor John Sterling, (whom the bigoted reviewers tried to make out an atheist,) somewhere says—

"To toil in tasks however mean,
For all we know of right or true,
In this alone our worth is seen—
For this we were ordained to do."[14]

{[F]inding a grove of suitable timber near the site of what was proposed to be Milwaukie bought cross cut saws, hired some men, put up a little shanty, with cooking stove and such appliances as would make a local habitation, we got to work. It was without doubt hard work for a greenhorn. A crosscut saw is intended for two men to work alternately, and cuts coming, and going. You pull with a rocking motion to make your cut, the other man pulls back and you should let him do it, simply holding the handle loosely. In your enthusiasm, you help the return motion; your friend at the other end does not find it necessary to restrain your heaven born energy, and until you "learn the ropes" you measure yourself up with your colleague, to whom this business is one of apparent ease, to your own great disadvantage.

{For a day or two you are too tired to eat; you are an employer, and it takes a great deal of tact and natural capacity to manage men who know all about the business of which you know nothing, and the cheapest way is to learn the business practically, at least to a degree, and then if you have more brains, you are in the lead without effort, if you haven't more brains, then the other fellow will soon hire you, which is carrying out the Lord's intention, I suppose.

{It was delightful being in the great woods, our selection of trees to fell was the reverse of logging in our own woods, here it was to eliminate the large timber. One of the clauses in the contract was that no log should measure over forty four inches, as their gates could not take a larger size. It was raining all the time, a steady drizzle which we could not escape during the day, but at night we were happy. The bark of the huge cedars stripped very readily, and strips ten feet long and three to six feet wide, made an impervious roof, and was easily got and applied. Built, say ten feet back of a fallen trunk five or six feet in diameter, against which is heaped a lot of the smaller limbs of the trees you have made logs of, you get a heat in your leanto that makes you regret you did not make your fire farther away. A substantial supper, a good bed of cedar twigs with a covering of fern, makes a bed which a king might envy and with your wet clothes drying finely on a pole stretched under your shelter, and the deep wood aglow with your cheerful and generous fire, and the pattering of the rain upon the roof, emphasizing the cosiness within, a song or two that goes cheerfully out into the darkness beyond the flickering gleams that come and go through the deep aisles of the listening woods, to where the shadows creep forward wistfully to the cheer which they may never know, a reminiscent story, of a mutual experience, and then "a sleep full of sweet dreams and health and quiet breathing."

{Down through aeons of ages, has at last come to us this boon of unfearing sleep, while all about us, what our blunted senses deems the sacred stillness of the woods, is vocal with horror. To all the small creatures of the forest the terror of the day has but given way to the dread of the night and every clump of grass, every sheltering shrub, is the theatre of a tragedy.

{Thank Heaven we cannot hear. It were to doubt the goodness of God, to believe they fully know.

{When we had got enough logs to justify, I hired oxen to haul them to the Willammette upon whose banks we were, and a boom held them in place untill we were ready to raft them.

{In a few weeks we made up our raft, and moved out into the current, where I took my first lessons in the movement of the tides; we moved off beautifully on what seemed sufficient current to take us down to Portland before nightfall, but grounding slightly on the head of a bar, we got over board to push off, and found it futile. In half an hour we were high and dry, and then the inland waterman remembered the tides, and knew he was settled until the tide came in and commenced ebbing. As we had no skiff, we were waterbound.

{Fortunately I had got my mail from Oregon City, and while the daylight lasted, had a delightful time reading over the letters.[15]

{The National Era, held the platform of the Liberty Party and my heart swelled as I read it. It was a bugle call, it seemed to hold the promise of a glorious future, and I felt renewed hope, for the nation. The contest had been long and the end was not yet, for the Liberty Party could not possibly hope for present success, but I felt as never before, that a cause so long sustained and so nobly presented must win at last and the curse of slavery come to an end.

{It was dawn when we reached the mill and got our logs in the boom and measured.}[16]

Decision Is Made to Move to Puget Sound

There is nothing mean in hewing out saw-logs and rafting them down, or indeed in any species of labor; but its deuced hard work, though, and was one of the resources for subsistence I least would have thought of ere I left home. It *paid*, however, even with all the necessary outlay in purchasing tools, opening roads, &c. The mill owners were anxious to give me a winter's contract, amounting to some thousands of dollars, and were much disappointed when I closed accounts, and announced my intention of leaving that part of Oregon. To this determination I had come for many reasons; the rainy season (a six weeks' incessant rain,) had set in, and though inured to wet and exposure, and hearty and robust to such an extent that you would hardly know me, I did not want to carry the experiment too far. A great many of the emigrants, too, by this time, had arrived, the majority of whom, together with a large number of the old settlers from the Willamette Valley, were going to Puget Sound in the spring. Hitherto public attention had been but little drawn to that part of the country, for the trains of the previous year had, (overjoyed at the transition from the dry plains and sterile hills they had crossed,) settled in the first valley they had come to, till the best of the lands in the lower country, especially the Columbia and Willamette Valleys, were taken up. At least 20,000 had started for Oregon alone this year, four thousand of whom, it is said, perished by the way-side. The remainder (16,000) would have to go to the mines, or else Northward to Puget Sound, and I was anxious to be ahead of them.

Besides this, I had heard of my cattle. When they were thrown on my hands after we left Fort Boise, I had engaged a young man to drive them on to Portland, but he had been unable to do so, on account of ferriage being so high, (crossing the Columbia alone costing $10 ahead,) so they were now at Vancouver.

I determined to go for them, and struck at once for Puget Sound, where I knew they would be a capital for me, commenced making immediate preparation, settled up all my "logging" accounts, purchased new outfit of clothes, &c.—While in Portland, I had met a friend, who had been up to look at the country and select a future home, was much pleased with it, and very anxious I should be a neighbor. "Stillacom"[17] was the place he had selected—a spot but recently laid out. The claim next to the city lot was not entered, nor the one above. I had a letter to the proprietor, which I knew would be well received, and without promising to accept any of the propositions from either him or my friend, I promised to call there on the way, as I had to go up to Oregon City for a letter I expected from home, and also to purchase some bran for the cattle; so I got a sack or two, and jumping into a borrowed skiff, rowed up to the rapids,[18] a mile below the city, and from there walked up, called at the mill, got my bran, and with it on my shoulder, and the expected letter in my bosom, I was threading my way, head down, and staggering under my load, when I was hailed by Governor Gaines, who came across the street, and said "I looked decidedly Oregonish." He seemed glad to hear I was going to Puget's Sound. Of the wonderful advantages of its extensive inland communication he seemed to have a profound conviction, saying a city would spring up somewhere there, which would be a Venice some day. He promised to give me several letters, but recommended me to stay till Spring, as it would be almost impossible to get the cattle over in such an inclement season, the winter being one of the coldest ever known in Oregon; but I was determined to go at once; log chopping had made me hardy, I felt equal to anything almost, and proceeded in my preparations. Wilson[19] (the friend spoken of,) agreed, for the use of a yoke of cattle, to purchase a wagon to carry what few personal effects I had, and also to pay the half of the expenses of the other cattle across the trail. If you will look at the map, you will see the course we proposed to go. He was to take the wagon, convey himself and wife from Portland, down the Columbia to the mouth of Cowlits on the steamer "Lot Whitcomb," and thence up Cowlits river to Wurbass Port,[20] where I would meet him with the cattle, and from there we would go to the Sound together.

Fort Vancouver and the Hudson's Bay Company

I started on Monday[21] over to Vancouver, where the cattle were—just twelve miles across. The road crossed the Willamette river at Portland, and struck straight across to the Columbia. Like all the roads in the country, it is but a "trail," and exceeding difficult to follow; one must rely more upon the direction he must follow than upon any indications furnished by traces. About the time I grew pretty thoroughly drenched by the falling rain, and began to think the twelve miles were the longest I had ever traveled, I came to a little clearing in the woods, where, to my great joy, I met the young fellow[22] spoken of in former letters, to whom I had been enabled to be a "friend in need." He was overjoyed to see me, and overwhelming in his professions of gratitude; cancelled the obligation, made me remain to dinner, and then accompanied me part of the way, ferrying me over a slough. I would have had difficulty in getting around. He was a blacksmith, had put up a neat little shop, was already "doing a good stroke o' work," he said, and evidently *was* on the high

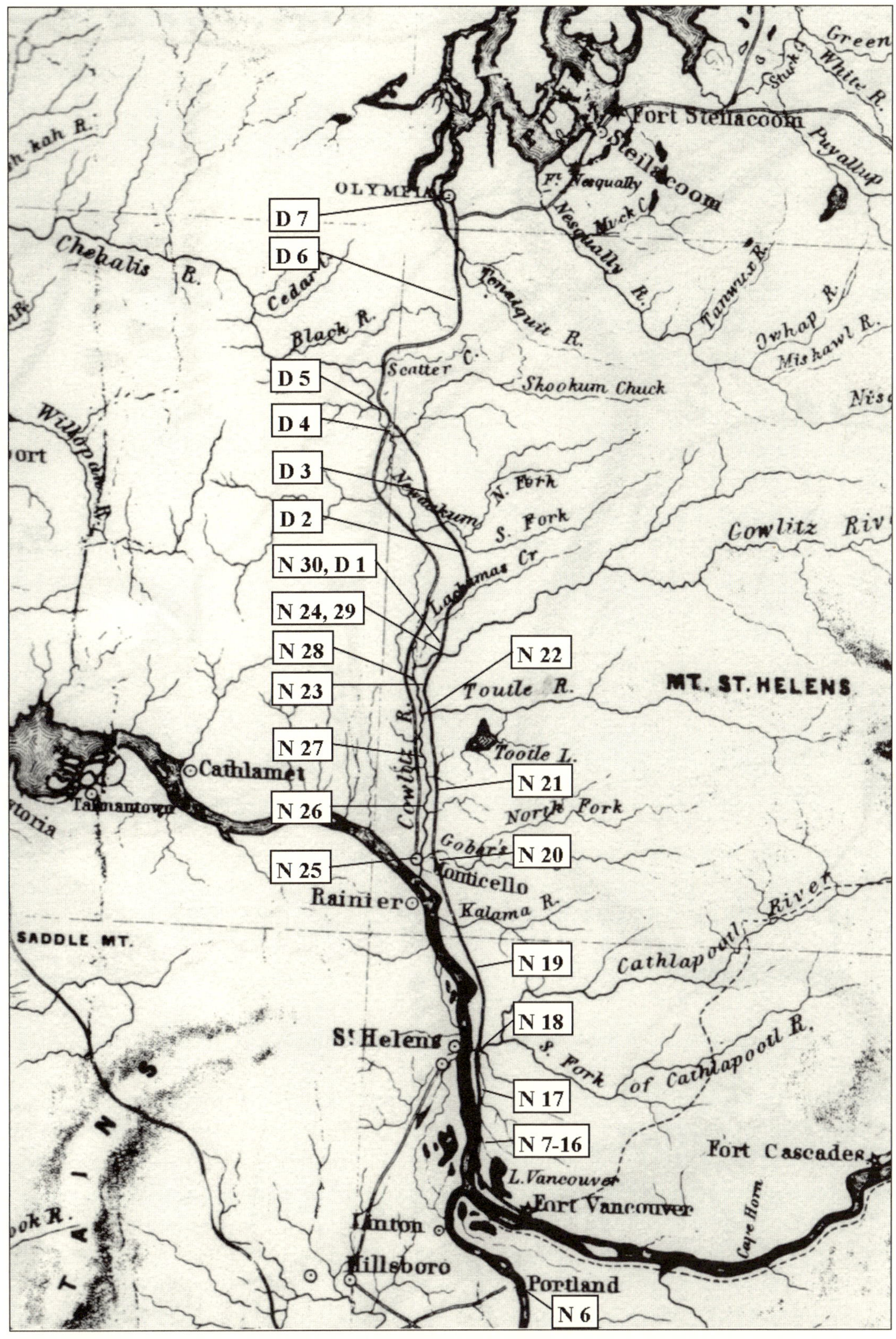

This map shows Allen's day-by-day progress northward on the Cowlitz Trail, which led from the Columbia River to Puget Sound. His journey began in Portland on November 6, 1853, and ended in Olympia on December 7, 1853. Boxes indicate Allen's travels in November (N) and December (D). Base map from Department of Oregon, Map of the State of Oregon and Washington Territory, Bureau of Topographic Engineers, 1859.

road to prosperity. I journeyed on two or three miles further, guessing as to the course, and at last emerged on the banks of the Columbia, but could see no sign of Vancouver, though the river up and down for two or three miles lay extended before me. Seeing a smoke some distance above, I went up and found a lodge of Indians, and asked for "a canoe, in which to cross the river to Vancouver?" but they had none but "a small canoe, and river too rough—wouldn't go." So I retraced my steps, and, following the river a mile or so, found a regular ferry, crossed over in a skiff, and found myself, then just in the "gloaming," minus one dollar, landing on the beautiful prairie on which stands Vancouver. After a moment's lapse I was going at a fast trot beyond the Fort, and soon after was in a comfortable little log house, beside a blazing fire, waiting impatiently for the supper preparing for me in the one-roomed "hotel" of the place. Supper over, we lay down to rest. My companion was a gaunt trapper, with no control over his limbs when asleep, and I dreamed of being hugged by a bear.

The Sabbath morning was ushered in clear, calm and beautiful, the sun shining down glowingly on the broad prairie, causing the grass to glisten as if strung with diamonds, making the walls of the fort look doubly grim by the contrast. It is a picturesque old place, though, and I made a sketch of it, (as I have done of all noteworthy objects in my route,)[23] which some day I will send you. The fort stands at some distance from the village, is built of upright pickets twenty-five feet high, enclosing about four acres of ground, comprising in its limits the houses, shops and magazines of the once powerful Hudson's Bay Company. Within the stockade reigned the agent or chief packer[24] of all the Northwest or Columbia department, including Caledonia,[25] with all the power and pomp of a feudal baron—extending to all travelers of "consideration" that unbounded hospitality so characteristic, which has elicited so many encomiums from persons partaking of it, and has, in too many cases, blinded them to the iniquitous policy they have always pursued to their subordinates and others in their power. To their credit let it be said that they have always, when in the zenith of their power, extended equal toleration to *all* missionaries, of whatever sect, and have always exerted their influence against the sale of spiritous liquors, or its introduction into the country; but the Indians have not improved under their sway; and their "employees," by a system peculiar to themselves, but characteristic, I am sorry to say, of old Mother England's Colonial System, wherever extended, have degenerated into mere vassals of the company. . . .[26]

For a great many years the company have had exclusive possession of this beautiful place, but have dwindled into insignificance—removing their strength North. The soldiers belonging to the station live in log cabins, built with some degree of regularity, on the prairie—many of them having squaws for wives, and living in a God-forsaken manner, characteristic of all such unions. The officers live luxuriously, and are generally very unpopular among the settlers, on account of their reputed arrogance.

Chasing Cattle

Adjacent to the fort and scattered in the neighborhood were a great many of the emigrants who had come over, and were here to winter. Among them were nearly all of the train I had come over with, as far as the crossing of the Lewis river.—Next morning, the first house I entered was Mrs. Hale's,[27] who seemed very glad to see me, and gave me information where to find the rest of the party; following which directions I went down the prairie four or five miles, and after some inquiry found Martin Kounts,[28] and with him the young man who had charge of the cattle. Anxious to lose no time, for fear the rains should make the roads impassable, we went out together to hunt up the oxen, and had a long hunt, walking at least twelve miles, and finding but one of them. Came home after dark, fatigued and drenched through. Kounts had split up some cedar, and had made a pretty nice little shanty. Here, by a warm fire, we sat until very late, rehearsing the "incidents by flood," occurring during our passage down Lewis river together. Monday again saw us out on our hunting, and the evening saw us returning again, "wearied and way-worn"—and "ray-ther dumpish," from the steady days rain we had been walking in—and finding but one more ox. The prairie was some three or four miles wide, from the river to the pine bluff, and most curiously cut up by "slues," lakes, swamps, &c. So that we might walk a whole day in a mile square.—These lakes are filled with wild goose, ducks, seagulls, cranes, heron, &c.—and every day for eight days, we had the felicity of walking up and down these places, under an incessant rain—with the consciousness that each day rendered our future journey more fatiguing.

{The fatigue of one day was the history of the next, and for eight days we made these tramps, until we had waded and swam every slough, lake, and swamp within a radius of ten miles and succeding in rounding up 4 of the cattle. It rained every day, and all the day, during that period and we had the knowledge that each day of such weather made the road ahead of us still more difficult, and at the last the rain was falling with undiminished perseverance. It was evident it was filling an annual contract, and was doing it on the same old lines, and felt at home in the business. Madame Kountz was kept busy drying out suits of clothes during the day that we might have a comfortable suit at night, one's luxurious habits will cling to him even to the margin of effeminacy. We made no protest against sitting in our wet suits to eat our midday meal, in case we were fortunate enough to get in for it, which was seldom, because ten minutes afterward we would be as wet as ever, and the intermediate dryness was a drag on subsequent effort, but we clung in a weak way to the habit of sitting in the evening in an arid condition. It is difficult to at once change all old methods "Mutanter coelum, non animum qui trans current mare."[29]}[30]

Starting Up the Cowlitz Trail

On the ninth day we had found but five, and I was so anxious to go that I started, with the young fellow with me to drive, concluding to let the other ox remain till spring. Kounts wanted me to take him along also, but I persuaded him and his wife to remain till spring, promising to write in the meanwhile. All the "old heads" advised me also to wait till spring, representing the almost impossibility of making it in the winter; but I had promised Wilson I would meet him, and *must* go. So we started with our five oxen, raining meanwhile most dismally. Twelve miles out on our way, we were fortunate enough to come across our own remaining ox, and so with our lot complete went on our way rejoicing.

{In his new sense of freedom he had endeavored to secure its continuance by getting into "no mans land" and ordinarily his judgment would have been good. We lost a great deal of time in inducing him to change his views as to his winter's residence, and not even his old comrades society reconciled him to accept willingly our conclusions as to his future. I can conceive of no greater proof of the wisdom of the Creator than in the mental limitations of the lower animals. If this reluctant ox had but a little more memory, or powers of induction therfrom, we would never gotten him a rod down the trail, but he was'nt a logician.}[31]

We lost a good deal of time, however, in catching him, and only made twelve miles in consequence that day, stopping at night with an old resident, who told me many tales of the first settlement of the country and the grinding oppression of the Hudson's Bay Company.

Swamps, Mud and Endless Rain

Winding all that day through the woods and the wet underbrush, following a trail not as plain as a cow path, off which we wandered twenty times a day, having frequently to brush the moss off the pines to hunt for the old "blazers," was no cheering work. Neither was it any improvement to wade all next day through a swampy prairie, (made so by a constant three weeks rain,) where it seemed one vast lake, the water coming at all times nearly boot-top, and obliterating all traces of our road. We strove to keep a direct course over the prairie, but when reached the opposite wood found we had lost all traces of the road, and were compelled to skirt all round the swamp in search of it, till we at last found the opening, not far from where we had entered. Thus wading four or five miles across and around the swamp, we continued moving on briskly, in order to keep ourselves sufficiently warm to reach a house, and in blissful ignorance whether or not we should find one at all.

After many windings through thickets and swamps—passing on our way several Indian burying grounds—we came at last to the banks of Lewis river,[32] which had been swelled by the rains to an alarming volume. After hallooing for some time two men came over in a yawl, and told us we would have to wait some days for the river to fall, as it would be impossible to cross our oxen over, the river really dashing by like a mill-race. I was very anxious for a trial, but he refused to risk it. Being determined to proceed, I asked what he would charge to tie a rope to the horns of our leading ox—and going over in the yawl, just row as I ordered—$1.50, he said. I took the weakest ox, fastened the rope and started off. When they had advanced a few feet from shore I drove the others in. The men kept calling out and warning me that the cattle would drown, and to desist; but our animals had become strong from their rest at Vancouver, and were willing; I had been with them so long I knew what was in them, and persisted. The men kept rowing up stream, and finally all landed safely on the other shore. I had taken the precaution to leave an extra rope, that, in case any one of them gave out, we could assist them, but there was no occasion—all came safely over. The ferryman, during the operation, was white as a sheet, fearing we should lose them—said never since he had kept the ferry had any cattle crossed at such a stage of water. He offered us $150 a pair for them. But I would not have sold them at any price, for they constituted, as with the patriarchs of old, (except the man and maid servants,) all our worldly wealth. We remained with these men all night, and learned from them that there were four or five men ahead about a mile, who had come to a halt, finding it impossible to proceed further, and they assured us we would be compelled also to wait there till spring. Some ominous twitchings in my ancle almost pursuaded me to remain, but I was anxous to fulfil *my* part of the contract with Wilson, and not disappoint him, and therefore pushed on.

We soon came up to the party, and ascertained from one of them that he had been ahead two miles to reconnoitre, and found it so exceedingly swampy that it was impossible to proceed. His mule had mired twice, he said, and altogether told us a most dismal, disheartening story. Still we plodded on dogedly, and when we came to the place spoken of, found he had not exaggerated its difficulties, "nothing extenuated, nor set down aught in malice." The trail, in fact, that we were following, was so faint, that had we not found the swamp, we should have concluded we were on the wrong road. Through it we went, hardly knowing whether we were right or wrong, for there was no way of telling—pushing on keenly, to keep the cattle from miring, which threatened to occur every now and then.

{Had they been mules we would never have got them through; here is the advantage of the double hoof, it plunges into the mud making a big hole, when the hoof is pulled out, the hoofs close together and take less room than when they went in.}[33]

Several times my boots were pulled off, and I had great difficulty in recovering them again. We had intended making ten miles at least, as we started at day break, but after splurging in and round that swamp all day, when, after dark, we reached a house, we found we had only made six miles! Now when you consider that in ordinary roads we should have made 25 miles at least, you may judge of the state of this road. The good people of the house were astonished as well as overjoyed to see us, as they had not expected to see the face of living man again till Spring. Next day we varied the performance by the absurd movement of going five or six miles on our course and, concluding we had gone wrong, retracing our steps, only to be assured we had been *right.* Pressed on till we came to Pretty Girl river,[34] (so named by some amorous and susceptible trapper, who swore he saw an angel standing on the bank, dressed in white.) Finding a canoe here, (no girl—*she* had been carried off, a La Sabine, as a waif or estray, by the enamoured trapper,) we crossed over, encountered another stretch of low, swampy prairie, over-flooded by the rains, and over a spur of steep mountain, coming down the opposite side on another prairie, and after a long two hour's walking, came to a deep slough, which we had to wade. It was icy cold, and deep enough almost to require swimming at times, when the under current nearly swept us off our feet. It was chilling work, and after we crossed we had to drive full five miles more.—The oxen, being tired, could not be driven fast, which gave full opportunity for us to *feel* its chill. We came then to another slough, and hearing the deep gutturals of Indians on the other side, hailed them to come with a canoe to take us over, the cattle wading. We went a mile further and came to house occupied by a Kentuckian,[35] who was half drunk, very verbose, and excessively polite.

An Unpleasant Experience with a Frenchman

The fire was down, and we, nearly frozen, were trying to get up a blaze. Meanwhile our Frenchman[36] was hopping around like a monkey, executing a variety of inconceivable antics, under the influence of the liquor he had evidently partaken freely of—the most sensible of which was his bringing us some excellent brandy, which, half chilled as we were, was accepted most gratefully, and paying but little attention to the maudlin courtesy which induced him to present it *on one knee*. He had a squaw wife, and half a score of little children—half breeds—who were swarming all round the lodge. Hungry as we were, the cooking seemed filthy beyond degree, and we had become disgusted with the Frenchman, whose drunkenness had by this time assumed another phase—abusing his wife, and eventually thrusting her out of doors. We both itched to be "into" the fellow, but his fire was a cheerful one, burning brightly now—outside looked chill and cold, and we concluded to be prudent and remain quiet, and did so till he commenced beating his children, when I could stand it no longer, and jumping up, commanded him to stop—he had done too much. He *did* cease instantly, but told me if I did not like it I could sleep in "the Bush!" The Bush! well it was better to do that than remain near such a beast. I thought for a moment, as little crapeau flew to the poker, that we were going to have it "up and down," and in such a case, there would have been but one result—an uncomfortable and sudden exit of the little Frenchman from his own domicile. I was putting on my coat very deliberately, paying no attention to his vapourings, except to advise him "to keep a sharp look out, or I would make a cold Frenchman of him!" the singularity of which threat, and the tone in which it was uttered, causing him to cool down instantly.

Sleeping in the Rain

We had had enough of the place, however, and as he still continued drinking very freely, I knew we would "come together" before morning, and so left his house and out in the pitch dark we went, crossed the prairie, found a little thicket, lit a blazing fire with the Frenchman's fence-rails, and making a kind of wigwam out of the snow or rather "wax-drop" bushes,[37] which are indigenous here, and cutting some smaller ones for our bed, managed to sleep till morning, though sleep had to be wooed under disadvantageous circumstances, as we were wet through from our wading, the rain was continually putting our fire out, and then, too, our having eaten nothing since breakfast did not tend to enliven matters any.

{For your enlightenment, if ever it should come to you to camp out in the rain, let me inform you that you *can* sleep, if previous conditions demand it, even wet through, with the same, old rain that "did you dirt" still beating down on you, but you can't slumber with the rain beating in your face; take my word for it, and waste no time experimenting, but at once do as we did, spread out your slouch hat to its fullest, and with a deft use of twigs, platform it over your countenance and sleep like an infant. My companion whom I regret to say sometimes fails to see the humor in some of our situations, said he never did have confidence in a man who brought you brandy on one knee; having had no previous experience with men who transported liquor in that way, I am unable to say how much it might be taken as an index to character. That this fellow followed that performance with dastard action, is not conclusive. It is unsafe to generalise from a single fact.

{The young man was disposed to be very contentious about it, I began myself to feel a disposition to take this, or any other little incident very seriously, and knew that I should look upon all mankind, including this drunken Frenchman with more kindly eyes, if I could only "tank up" with a good breakfast. We are weak creatures. Bowels of compassion" seem to be largely contingent on a full stomach.}[38]

Next morning we started across the Cowlits to get breakfast. There were two or three Indians on the creek, and one of them we got to row us over. These Indians build much more permanently than any I have yet seen, and in a rude way pay some attention to comfort. They pick up slabs in the river, and setting them up on end, about five feet high, run girders across, and upon them erect rafters, just as we do, shingling with bark, which they cut from the cedar trees. The roof is open in the center, and in the middle of the floor they build a small fire, around which they all squat, while the smoke eddying up through the rafters answers the double purpose of curing the salmon hanging there and warming the house. A number of families will live in one lodge, and a dozen fires be lighted, each surrounded by its own circle. They manufacture of rushes a soft kind of mat, which they use to lie upon, and also to hang up on the sides of their lodges, to keep off the wind. They are not pleased to have any one come inside, but if you persist you might stand there all day and be unspoken to, not one of them paying you any attention or heeding your presence. They are much more industrious than the generality of the tribes, and do all the boating on the river—no inconsiderable amount. We got one of them to cross us over in a canoe, and in the flourishing town of Monticello, consisting of four houses, we ate our breakfast, and listened to the oft repeated tale of the horrible prospect we had ahead of us.

On our return the trail was very difficult to find, and our friend the Frenchman tried to put us on the wrong scent, in return for which I summed up all the French I was master of, to call him "Le Grand Liar." There had been a drove of sheep started from Vancouver nearly a month before, and though the weather was fine, they had only got two days in advance of us, and their trail being pretty fresh, we had but little difficulty in keeping it.[39] We struck the Cowbits [Cowlitz] river about six miles above, and passing by a clearing, were hailed by its owner, who pressed us so heartily to come in and eat, that we did so. He was a jovial, kind-hearted Irishman, and amused us not a little by his skill in turning flap-jacks, in which he took great pride, and *we* a warm interest. I had meanwhile been drying my clothes by his huge, blazing fire, and enlivened by the idea of two meals a day—a circumstance that had not happened before since our starting—felt equal to more tugging. We had a tremendous quantity of swamps to wade, creeks to cross, thickets to go through, &c., and at night encamped in the midst of a dense pine forest on a mountain top. There was nothing for the cattle to eat, and for fear they would take the back trail, we slept right across it, the underbrush being so thick on both sides that they could not possibly go back, unless over us. Cutting down a great quantity of young pine shrubs we made a nice little "wickeyup," and with hemlock branches strewn on the floor, went to bed, first cutting enough wood to keep up a blazing fire all night, and eating a while at a tremendous Ruta Baga turnip Wilson had bought from the Indians.

Our blazing fire attracted the attention of some Indians below, who had landed a canoe on the bank of the river and camped for the night, one of whom came up to us, and after devouring the remainder of the turnip, wanted us to go down and he would "Potack hivo muck-a-muck,"[40] (give us plenty to eat.) We went down, and found quite a number of Indians, with any quantity of dried Salmon, dried Raspberries, &c., bought a tremendous salmon from them, and having cooked and eaten it, a second time went to bed. I laid a long while awake, watching the flickering light radiating through the vast forest, making a halo of brightness in a field of darkness,—tinting the dark brown trunks of the pine trees, standing like columns around, through whose branches the stars could be seen faintly glimmering. I fancied the wind, murmuring through the tall pines, to be voices whispering to me of home and the dwellers there.

Dinner at Hardbread's

{Next evening found us at the fork of the Cowlits[41] where was one white man's house, and a great number of Indians. We had carried off some dried salmon from the indian[42] camp, but we could not possibly keep anything dry and while wet salmon *can* be eaten, it ceases to be *dried* salmon and is not palatable; so we were rejoiced to eat white mens food, and sit under a roof to eat it. The old chap who kept the shack was certainly an odd genius and had been in all kinds of places and had all kinds of experiences; cooking is not a fine art in this neck of the woods and toleration is a necessary virtue in a guest, but even so, this old man drafted to exhaustion on that virtue; he could'nt cook for an insane asylum.

{It was a luxury to have a piece of soap and a basin to wash in, and though the soap smelt unpleasantly, and was averse to yielding its grip on the cuticle, yet it had the suggestion of civilization, and we appreciated that. Our meal seemed somewhat delayed, but not from any remissness on the part of our host, the delay was ours, as we discovered when our salmon and potatoes were served to us in the basin we had been using; it seems there was a scarcity of utensils in the establishment, and the basin had to serve many purposes. It might have been well to have cleansed it after each diverse use, but the old man was a little careless in such matters.

{That soap, whether from our, or many previous uses, had a wonderful affinity for any thing it came in contact with, and seriously affected the flavor of the viands. It did not matter so much for the first few minutes, but became serious later. Somewhere in the old mans wanderings he had imbibed some method of trifling with flour, and producing some result which he called bread, and he fed this to us who were not in stomachic condition to refuse an edible compound of any nature. When a few days later on our way to Olympia, we stopped at Jacksons where a number of people were weatherbound, I told of my experience at this house of call, designating the host as "old man hardbread." The name was deemed fit, I think no one in the territory ever knew him by any other name. Years after in Ezra Meekers interesting story of early Oregon he is spoken of as Mr. Hardbread.}[43]

Woke up in the morning—cattle gone, could find them no where, nor discover the direction they had taken; the rain all night had obliterated all traces. Set an old Indian on the hunt, and found to our great dismay they had crossed the river and had gone back. Hired an Indian and canoe and went down the river two or three miles, intending to land and examine the trail, to see if they *had* passed down, and if they had not to follow up and drive them back. Six miles down saw three of them standing on the bank, having again crossed, while the others had gone down. Drove them over, hunted the others up, and made quick time to our camping place again.

We had caught the men who were driving the sheep at the Forks, but they had again got a day's start on us. Pushed on, and after a quick day's tramp caught up again at night. We camped together, under some gigantic cedars, somewhere about ten feet in diameter, and with dense branches, so interlaced that they kept the rain from us.

{The Scotchman[44] in charge of the party had ordered a sheep killed. We cooked the strips of mutton on twigs stuck in the ground before the fire, "every man for himself" and no limit either as to mutton or time. It was a perfect meal both as to quantity and quality. No sheep could in reason hope to leave this world with more appreciation.}[45]

There were some twelve or fifteen men in company with him, assisting in driving. It was a picturesque sight to see them all gathered around the camp fire; the majority of them being half breeds, (who, of all people I ever saw, are most fond of brilliant colors in their dress,) seated on great branches of fern that grows hereabouts, with their bright red blankets, crimson sashes and variegated leggings, their sparkling, vivacious eyes, long black hair, and caps set jauntily on one side, they presented a picture 'neath those dark old woods, worthy the pencil of Salvator Rosa [Italian painter (1615–1673)]—their animation presenting a striking contrast to the subdued, quiescent look and attitude of the "stoics of the woods," grouped around in the gloom like statues. Apart from all stood two "Kamakers"[46] or Sandwich Islanders, as fine looking men as I ever saw, and who, of a kind between Indians and white men, class with neither. All these grades are very tenacious of caste, believe in "priority and degrees" and never intrude upon each other. We Saxons ate first, then the "Kamakers" and half breeds, and then the Indians. Our sleeping part of the performance was very ill done—camping out in November is no joke.

The next day we passed through numerous Indian grave yards. The dead are laid in canoes, which are elevated on stakes raised a few feet from the ground. Holes are almost invariably bored in the bottoms of the canoe, whether to allow the rain to escape, or to render them so useless that they would not be stolen, I know not; all the effects of the deceased are laid beside them, their blanket thrown over and drooping from them like a shroud, but torn into shreds, lest the passing "Christian" might rob the grave. We passed through one very large place of burial, and among the number I noticed small canoes, evidently the last resting place of children. Some had rotted from their pedestals, and the mingled bones, sculls, &c., reminded me of the Scriptural Golgotha—"the place of skulls." I was struck with the formation of the skulls we examined. They were of the "Flat Head" tribe, who have a singular custom of flattening their heads in childhood, by compression {a board is bound over the forehead, and kept there until the head is permanently misshapen}. It would puzzle a phrenologist to discover any organs on these skull cases. The shape of their heads is very advantageous, however, to the squaws, on whom devolves all the carrying of burdens. They have a strap of wicker work, which is passed over the top of their heads, attached to a basket of the same material on their backs. When leaning a little forward, it cannot possibly slip. All in the tribe are thus deformed from infancy, except those taken in battle, belonging to other tribes. The natural formation of the head is considered a sign of abasement and servitude—the deformity an evidence of claim to be a "Tice," or gentleman Indian.

There is one feature in the burial service of the tribes of Indians I have seen, that seems to be general and common to all—a desire to dispose of all the deceased's property at death. A similar disposition in civilized countries, if prevalent, would have deprived us of reading "Bleak House," and saved much exasperation against the "Big Wigs."

Monticello Convention

That evening brought us to the termination of our journey on the Pack Trail. Crossing the river, we found ourselves in the little town of Warbose Port [Warbassport], where Wilson was to have met me with the wagon and go on with me. I met him, but he had not brought the wagon. It had lain for a week at Monticello. Finch,[47] the boat owner, could not hire any Indians to go down with him, as it had been and still was raining disagreeably and constantly. I put the cattle in a little prairie near. Warbose Post is a little place of two or three houses, not yet laid down on any map; by looking at Mitchell's,[48] however, you will see the Cowlits farm[49] laid down, and it is one or two miles above.

In the evening, after putting the cattle in good grass, I went over to the tavern[50] in the place, and sat looking over some old papers, by the cheerful fire. Old as they were, and of diver[se] politics, (there was one entitled the "*The Tocsin of Liberty, or the Tin Trumpet of Democracy*," edited by H. Mealy Squirt,—"a trenchant upholder of the principles of '98") they were interesting to me, and absorbed in their contents, I was startled by a slap on my shoulder that, in my debilitated condition ere I started, would have heaved me over; and looking up, saw Quincy Brooks![51] I was as glad to meet a townsman as he to meet one so recently from home, who could tell him news of his friends. As I rose up and straightened out to return his greeting, he seemed surprised—would have hardly known me, had he not heard I was here—so much increased in girth and stature, so healthy looking and hale, so *improved* in every way. I modestly received the compliments, endeavoring to waive them by asking a hundred other questions, but he returned to the exclamations of surprise at the transformation, till, excited by his wonder, I began to draw breaths like a porpoise, feeling my muscles, expanding my chest, and wondering how a small man felt.

He had come down with a party of delegates from the South [Sound], to attend a convention of the territory at Monticello, to memorialize Congress for the partition of the territory into Northern and Southern Oregon. On my first arrival in Portland, I had (having formed a resolution before I started that, health regained, I would settle on the Sound,) forwarded what letters I had from Washington city,[52] and had afterwards sent on those the kindness of Gov. Gaines had furnished me, together with a letter announcing my intention of coming in immediately; stating also my probable time of starting, and the route I would pursue. These letters were of service, as making me known more extensively and favorably than I had anticipated. By these means and the kind offices of Quincy Brooks and other friends, I had been elected a delegate from a district on Budd's Inlet, and they were looking for me all the way down. Here was a transition—from "bull driving" to politics. Of course I couldn't deprive the territory of my most valuable services in *such* a "crisis," and abandoned all idea of further progression on my journey till the convention had adjourned.

They were going down in canoes, and I went with them, and attended all the sittings of the convention, the proceedings of which you have no doubt ere this read. That convention was composed of as good material as ever I saw on any public occasion—a body of earnest thinking men, who knew what they met for, and the readiest way of obtaining it. What it lacked in refinement and dignity was more than compensated by the good sense every one seemed to possess, and although but few lawyers and politicians were among them as "speech makers," you must give them credit for tact, shrewdness and political sagacity, in so timing their memorial (which, if granted, would bestow on the Presdent elect a new batch of offices wherewith to feed the cormorant demands of the army of office-hunters beseiging Washington city,) that there could be no fear of its rejection. The result was as anticipated. The memorial and petition (which I had the honor to sign,) was granted and adopted with but (I think,) a single alteration—the substitution of "Washington Territory" for the new country, instead of Columbia, as we had suggested, and which change was a most excellent and appropriate one. My part of the proceedings of that convention may be briefly stated. "Unaccustomed as I was to public speaking," I *said* but little. Excited by the novelty of my position, I *did* feel as if I could have been eloquent, and commenced by "taking a retrospective view of all the nations and empires that have ever flourished;" but as this somewhat comprehensive survey seemed incompatible with the limited time before us, I was obliged to restrain the fervor and afflatus, and content myself with what service I could render in the writing department, with personal thanks to each member for the consideration shown in my election as one of themselves, and their courtesy and friendship since continuing—assuring each and all of them that "it was the happiest and proudest period of my life."

As I sat in that convention, taking notes of its proceedings, I could not but think of my unexpected position.

"Some men are born great,
Some achieve greatness,
And some have greatness thrust upon them."[53]

I was not *born* great, for according to all authentic accounts,

"Our old Saxon blood
Has flowed through Plebians ever since the flood;"[54]

I have not achieved any particular greatness, that I am aware of, unless a successful accomplishment of a journey through hardships and exposures under which many who seemed more capable than I of doing so, have failed, and whose bones lie bleaching on the plains, or mouldering in the wilderness—may be considered as such. Therefore must it be said that greatness has been thrust upon upon me. So be it—it came unsougt. But (modestly speaking) I trust I have "that within me" equal to all fortune's favors, and flatter myself that the dignity and importance of that convention, which had the honor to create a new Territory, some day to be a State rivalling in prosperity the proudest of our land, was not lessened by the deportment of its youngest member. There was an unofficial meeting after it had adjourned, in the room or loft of the only hotel in the place, where twelve or fifteen of us were crowded together to sleep, that it would have taken the brush of a Hogarth to paint, and the pen of a Dickens to report.[55] The convention was composed of all kinds of material—and it seemed as if in that room were congregated specimens of all minds, and natives of every country and clime. Most of them were middle aged men, who had breasted the stormiest of life's battles, and yet, in all and through all, had preserved the enamel of the heart, and all its freshness. Jovial and gay, too, they were, with a fund of health and flood of boisterous spirits that would have entitled them to an honorary membership in any branch of that honorable fraternity called the "Jolly Cocks." There were old employees of the Hudson Bay and North-West Company—hunters and trappers—who had tramped the mountains and plains "solitary and alone," from the Russian Possessions to California, and from the Cascades to the Rocky Mountains and *over*—some of them for forty years; farmers from Iowa, Wisconsin, and good old Pennsylvania; some of Doniphans'[56] men; old whalemen and ship captains, who had traded from Kennebec to Astoria. Each had his own true tale to tell, and very strange and marvelous were the stories told that evening; I listening in bed to the quaint narrations of ship-wrecks, Indian treachery, &c. They talked till it was near the "wee short hours ayant the twall,"[57] and seemed never fatigued nor disposed to cease. As the hours waned away, a more sombre feeling seemed gradually to brood over all; their thoughts seemed to revert to their youth, and reminiscences of boyhood were recalled and reciprocally related, with as much prolixity but more feeling than the relation of the later passages in their lives. There were some singers among them, and almost every one in the company had either "given his experience" or sung a song, until at last, when the conversation flagged, I was called upon.—Now, I'm not much of a singer, though, as the Irishman said, "as good a whistler as ever twirled a lip." But I essayed to comply, and thinking, in view of the state of feeling seeming to prevail, nothing could be more appropriate than the reminiscences of Mr. Benjamin Bolt's[58] boyhood, I "lifted up my voice" and sang it with such a "vim" as caused two old whalers to rise up from their beds and shake hands across their six intervening companions with a fervency that, like old Dietrick's smile at the Christmas dance at old Yardle's, "baffled all description." This incident gave a new turn to affairs, and an exciting discussion sprung up between two "old covies" respecting a particular feature in the ancient game of "mumble peg," to settle which they simultaneously jumped out of bed, "accoutred as they were," to decide it by going through the game "regular." At it they went the jack-knife flying all round the room during the process—now projected from tip of nose, top of head, chin—all in regular orthodox fashion; correcting each other's mistakes with all the enthusiasm of boys, while we, raised on our elbows watched the progress of the game with intense interest, shouting approbation to each scientific and skillful throw. It was a comical sight, and yet a beautiful one, to see those two old weather-beaten fellows, whose faces bore the impress of many a year's struggle, and whose hearts might have become hardened in the ordeal, thus absorbed in one of their boyish pastimes, and as intent upon the game as if the long line of years since they had played it had fallen away, and they were once more boys together.

Canoe Trip Up the Cowlitz River

The Convention over, like a true patriot, I resumed my journey, and no one would have supposed, from any unusual elation of spirits, or extra "putting on airs," that I had borne a part in its proceedings. I bore my "blushing honors" meekly, I do assure you. Most of the members from the Sound left the next morning early; I didn't get away till later. We had a tremendous big canoe, with a large load in it, and started in the same old rain, which had never ceased, nor evinced the least shadow of a sign that it *ever would*. The Cowlits river is about as large as the "Conemaugh,"[59] but deeper, though so full of rapids and falls as to make it almost unnavigable. We pulled hard all that day; I was earning $2.50[60] per day, besides accomplishing my own journey, but it was hard-earned money. We had passed the canoes carrying home the delegation, but they had soon overtaken us, shouting forth to hurry on to Olympia.

Next day it rained only at intervals, affording us glimpses of sunshine now and then; so we made better headway. We camped upon the shore, being two or three miles below the nearest house, and pitch dark. Our matches were wet, and as all the woods were wet also, our prospect of having a fire seemed slim; but we had been on the "trail" too long not to be fertile in resources for all exigencies. We cut a "jaggle" of a tremendous pine tree,[61] out of which the turpentine poured in a regular stream, loaded up our gun, and firing it bang into the turpentine, soon had a lively blaze, and from it made up a cheering fire. We had brought some "Napitoes" (Anglice, Potatoes) with us from the last house, and had some beef on hand, and a bag of shorts,[62] and out of these and the clear running water of the river I made *such* a pot-pie—"you never see," nor did we see it very long, either, for it soon disappeared.

Our next day's journey brought us to the Forks, our course leading through a region giving every evidence of having been subjected to volcanic influences. The strata all along the river is reversed, thrown up and aslant—while to confirm all, rises, some fifty miles to the right, Mt. St. Helens,[63] with an unextinguished crater still on its summit, though its sides are covered with the eternal snows. I saw in several places also what my little geological knowledge assured me were signs of coal. Above the Forks half a mile, *is* a coal mine, owned by a California company, who intend taking the coal down to the Columbia, and thence to San Francisco.

At the Forks I met a gentleman who was one of a party who went on an exploring expedition up to St. Helens, and gave a glowing description of the wild scenery they passed through. That range of mountains I have no doubt contains much gold and silver. The Indians upon the Columbia bottoms go up there in the summer and bring back much of it, but refuse to tell the locality, as their chiefs have forbidden them, fearing the whites will do as they have done in Shasta—take up all their lands. I doubt not but all that range will be found full of gold. It will probably become developed just at the time when the country becomes settled, and there will be enough produced in the country to support the universe.

The company who went out prospected slightly and hastily, and found some small quantities on the river bars. They were too late in the season to prosecute their search as far as they wished, and intend going again next spring. They invited me to join them. I should have liked to have done so, had I had time; but my object was to take a canoe load of provisions, go down the Sound, make a complete exploration of it, and search for a good locality to repay my toils and time—a good water power would be invaluable, and there are doubtless many yet unknown. However, of my future plans more anon, in their proper place.

I had a narrow escape that night. I had made a kind of wigwam, and covered it with a wagon sheet to fend off the rain. In the dead of night, a dead branch fell from a perpendicular height of 100 feet, and lit upon my hut, breaking it down, and "fetching" me such a pelt upon my side as "brought me up standing." Had it not been for the ridge pole of my tent, which broke the force of the fall, I should have been killed, as sure as fate.

We fished up an Indian next day, and with his extra help managed to reach Warbassport (the place I had turned back from to "serve my country") late at night. Still raining, when I tumbled into bed and slept as I have learned to sleep in this country—sound as a top, yet wakeful at a whisper.

It took us all next day to fit up our wagon, and then our cattle couldn't be found and we had a hunt through the wet underbrush for them. We were three days finding them, six miles from where we left them. I trailed them three miles, and then losing it started an Indian off to hunt them. I verily believe an Indian can *smell* a trail—at all events, he found and brought them to us, and on Saturday we were off again.

A wagon had started off the evening previous, we were bound to catch them, and the style in which we put our cattle through wasn't slow. They had been on splendid grass, we had also given them a liberal feed of hay, and they were in prime condition.

Jackson, Borst and Ford

We reached Jackson's[64] that night just one hour after our friends. Jackson was one of the Delegates, and treated us bravely—for the first time since I left St. Louis, sitting down to a home-like meal, good biscuit, roast pork, fresh butter, milk, and potatoes. He told us to "go in and win." We *did.*

I bought a cat at the Forks, from an old man there;[65] I don't know why I did it; as a general thing I dislike cats. I wanted to have something in my new home attached to me, and something of originality about this specimen prompted me to buy it. I took it in the canoe to Warbassport, and from thence in the wagon. It was an eccentric kind of an animal. Beside other performances I need not particularly mention, it evinced a great desire to end its days by jumping out of the wagon, when it hung by the string attached to it. I cured it of its suicidal predilections by allowing it to hang there awhile.—This little investment proved a profitable one in the end.

Ugh! I feel chilly even now, when I think of our next day's experience. We came to the little river "Nowagwam,"[66] and in driving in struck the wrong ford, and got in so deep that the water ran in over the top of my wagon bed, and our cattle got so tangled up in the brush that I had to jump out and unhitch all but one yoke from the wagon. It was chillingly cold, and the water very deep and swift. Once or twice I lost my footing, and had to swim for it. We were walloping about in it at least twenty minutes, and shook, when we came out, as if under a fit of ague. I managed to get the wagon out safe with but one yoke, but we were carried by the swiftness of the current far down the stream, and I expected no less than to drown the cattle, but we came out safely. Our neighbors, hitting the right place, had less trouble. Wilson's wife, who was in our wagon, added to our anxiety and fear, but when all were safe over, bless her womanly heart, she compounded a mixture for us, of brandy and camphor, that warmed our "witals," and no doubt saved us much after suffering. Worse than all, after we were over, our course led through a prairie up to knees in water, and long after night we reached a house, where we halted. Every thing in the wagon was wet; I was a "deuced uncomfortable moist body" myself—and had never felt so completely done up, since we left the "Main Trail." To rhnder [render] matters more gloomy, the cabin contained but a thimble like stove, and a dozen were seated around it. There was but one resource; I compounded a kind of stew, so hot that in any ordinary trim it would have taken away my breath, wearily threw myself down in a corner, and arose in the morning good as ever, except a feeling of weakness and lassitude. Just as we were starting off, our neighbor, whom we had overtaken, came up without his wagon, which he had left three miles behind, and had given up in despair. We reluctantly bade him farewell, and pushed on alone.

The next wood was three miles through, and it was mud and water up "leg high!" I never saw or could imagine anything equal to it.

{Certainly to be called a road, it demanded more toleration than the average traveller has on tap. To make mud to that extent, there must be a good depth of soil. This fact did not seem to bring much hope or comfort to us, and none at all to the cattle.}[67]

The poor cattle began to fag out, though we had scarcely any load. It was, as a Western orator said, "awful! gentlemen, awful!" Then a prairie again, water six inches deep all over it, looking like an extensive lake. Then passed some more Indian villages, on higher ground. Then a deep slough, which we had to ford, breast high, which brought us to the banks of the stream we dreaded most—"Scukum Creek"[68]—which is the Indian name for "Swift-water;" but here, to our great surprise and joy, it was not as bad as it was represented. The Indians wanted to cross our cattle, but I drove them in, and they crossed themselves. The Indians, however, took the wagon over, with Mrs. Wilson in it, and all were over safely, including the cat, who seemed to enjoy the ride—and once over, our journey for the day was over. We lodged in Joe Boist's[69] house—a generous-hearted old settler, who amused us, as we all, after supper, sat round his cheerful fire, by his narrations of his struggling years, Indian customs, &c.

The Skukum creek empties into the Chickalees [Chehalis][70] river just here. The Chickalees is a lovely little stream, which empties into Gray's Harbor on the Pacific coast, and flows through a most rich and beautiful country, very little known and settled as yet, but which will, in my opinion, one day prove a most productive section. It possesses, from report, a safer harbor than that at the mouth of Columbia. Its headwaters swarm with salmon, and from its mouth up, it is navigable for ships—how far up I do not know, as I do not believe it has ever been explored. There is abundance of good soil and timbered land on its banks, and the country is prolific of game—elk, &c

From "Nowagwam" the prairie became gradually more and more gravelly, till when crossing "Shukum" creek it becomes quite so, and retains no water. It is however, an excellent grazing country, and raises as much wheat and as good as the better lands. I was passing over it at the most unfavorable period, in the dead of winter, and during the time of the annual rains, which, like the overflowing of the Nile, make fertile the land; but are also the only drawback to the pleasantness of the country. We were grateful partakers of old "Joe Boists's" hospitality for a whole day, taking that opportunity to give our cattle a respite, and wash and dry our own clothes. Toward evening hitching up and going three miles, reached "Ford's." Ford is another old settler.[71] Brooks had told him of my coming after, telling him to watch for me. Ford teased me to sing for him, and I am afraid I made the tender hearted old man melancholy all night, by butchering the beautiful air called the "Old Folks at Home." The hospitality which is characteristic of these pioneers is very grateful to the weary traveler, whose comfort depends much upon its exhibition; and I know of no way to elicit it more readily than by greeting them in the same frank and free manner they greet you—after the manner of Leigh Hunt's advice:

> "For wheresoe'er his own frank looks were thrown,
> He struck the looks he wished for with his own."[72]

This has been my course always, and beside being in consonance with a natural open and unsuspicious disposition, I find it good policy.—At parting, Ford would accept nothing for our lodging and fare, and gratefully we "went our way."

All next day we were traveling over beautiful plains, and one very singular one—all upheaved in little mounds[73] about five feet high. What was the origin of these mounds I know not; they are too high to be raised by the "Gophens," [gophers] or by any other animal I know of; besides there were no traces of the animal to be seen. The "Chickalees" river runs not far off, and it may be in remote time, has inundated this prairie, and in its whirlings worked out these hillocks; but why it had not the same effect upon the neighboring prairies, which are on the same general level, is also curious.

It seemed impossible we should have pleasant journeying any length of time, and we soon come to another slough, which we ought to have crossed with but little trouble, but our cattle, poor brutes, got

Mound Prairies, or Mima Mounds as they are known today, south of Olympia. From "The Mound Prairies of Washington Territory" by George Gibbs, *Frank Leslie's Popular Monthly*, 1883.

bewildered, stopped in the middle of the stream, and it required all our shouting and some little beating to induce them to proceed. It was necessary to jump in the water to do this, and for fear of getting chilled I drove them almost at a trot to the next house, three miles off. I came across a young fellow here, whose father I knew, having ferried him over Lewis river, at Fort Boise. He received us heartily, and gave us a good dinner.

Arrival at Olympia

The road from thence wound through a thicket so crooked that it was a wonder we got thro at all. The young fellow had said a wagon *might* go through, as one loaded with wheat *had*. "What man *has* done man *can* do," and we *did* get through at last. The man who laid out that road must have been gimlet-eyed.[74] After getting through this we fortunately encountered no more sloughs, and at evening arrived at a house wherein lived a blacksmith all the way from Maine, who had come 2,500 miles to get in precisely the same latitude of his native home. We had slightly broken our wagon coming through the wood, and remained to have it repaired. Our cattle next morning were hard to find, as there was but little grass on the prairie; but we found them at last, proceeded on our way, and early in the evening, in the "gloaming," found ourselves at "Newmarket,"[75] at the head of "Budd's Inlet"[76] and at the mouth of Shute[77] river, while stretched far down we could obtain glimpses of the "Sound." Two miles down the bay was Olympia, and jumping into a canoe and hiring the Indian who owned it to row me down, in a few minutes I stood on the beach at Olympia—having walked one hundred and fifty miles through an incessant rain, always wet through, and scarce half fed. This completed the last chapter of the somewhat "extenuated" but nought set down "in malice" description of the "Oregon Trail.

I stopped all night at the frame tavern dignified by the name of Olympia Hotel,[78] and in the morning made inquiries about Stillacom, where I had been asked to settle, and found there was some difficulty about the title, it once having been a post of the Hudson's Bay Company, and reserved by them in our treaty with England. So I concluded to go no farther, at present, feeling the necessity for a little respite from my arduous journeyings. Wilson, also, agreed to remain, as indeed he could not do otherwise, and

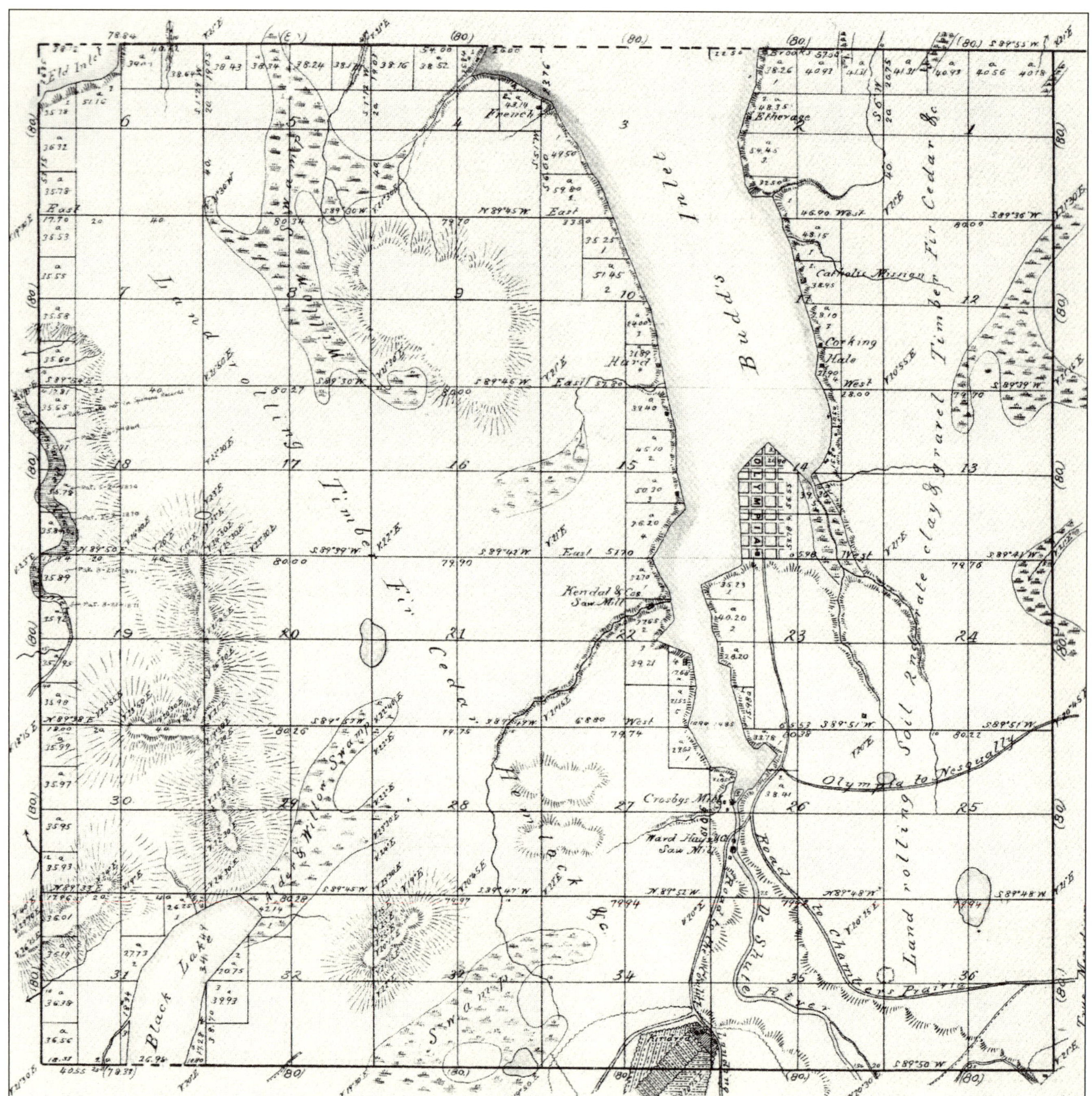

The Olympia area as Allen knew it; this map was drawn in 1853-1854. Map of Township 18 North, Range 2 West, Department of the Interior, General Land Office. Courtesy U.S. Bureau of Land Management.

we hired a little house, stowed in it what effects we had, and I thought for a time I would occupy it, but Brooks came up next day from his claim, two miles below, where he had some wood-choppers cutting timber, who paid for the privilege by boarding him, and invited me cordially to go with him, till I looked around—an invitation I frankly accepted. His house is built upon the shore of the Bay, which is here about three miles across, and is about thirty feet above high tide, the ground behind ascending gradually to a height of forty or fifty feet, before it reaches the level. This is the general height of the banks all the way down, though we are not upon the Sound proper, but the Inlet. I sat at the window of his little board cabin, with a glorious "outlook" from it before me, watching the long swells as they roll in upon the beach as the tide ebbs.

Description of the Territory

I wish I could give you an idea of the great extent and importance of the Strait of Juan Fuca, and the wonderful Bays, Inlets, Channels, Straits, and Harbors, forming that wonderful net-work of inland communication called Puget Sound. Northern Oregon or Washington Territory (as we have made it) possesses immeasurable advantages over Southern Oregon, in the superior harbors and means of ocean and inland intercommunication.

The Columbia bar we all know, is only passable in favorable weather, and dangerous at all times, under the most favorable circumstances—ships frequently having to lay off for subsidence of wind for weeks. The wreck of the "Peacock" is still fresh in our recollections, as also the report of Commander Wilkes, of the cause of their loss. No city up the Columbia, therefore, can ever be the New York of Oregon. From thence, along a wrecky, dangerous coast line of 700 miles, there are no harbors of any size or safety except Grays' Harbor, (which has never yet been explored,) till you come to the Strait of San Juan Facuba [Juan de Fuca], which is described by Wilkes as having a safe entrance, shores bold—anchorage found in a few places, wind generally blowing, sometimes severe, but always outward, and so deep that bottom could not be found in many places within a boat's length of shore, with a sixty fathom line. Very little is known generally of the extent and magnitude of these wonderful Inland Seas. They were never completely explored until it was

Olympia as it looked on July 3, 1857. Although made a few years after Allen's residency, this is the earliest known depiction of Olympia. Sketch by James M. Alden. Courtesy Washington State Historical Society, WSHS# 1932.93.17.

done under our flag, by Commander Wilkes, and his account is not generally read, the high price ($20) for the cheapest copy, preventing a general circulation. The Government copy, I believe, cost $125, and the members of Congress unanimously voted to each of themselves one or more copies. It is an excellent work, however, (all that part relating to Oregon, I know nothing personally about their experience in Nootka Sound and other interesting localities.) In addition to an Inland Expedition, they explored all the Strait, the myriads of Inlets, Bays, Canals and Ports, stretching from the Strait into the far heart of the country to Puget Sound; Ports Discovery, Townshend, Lawrence, in Admiralty Inlent; Hood's Canal and all the many and various passages of the land-locked ocean, washing the shores of as fertile a country as the average of our Southern soil, with the additional advantages of being a mineralogical and heavily timbered country with as equable a climate, though a little colder. He describes all these Inlets and Bays, as equally safe; the shores so bold that the ship's sides would frequently strike the shore ere her keel would touch bottom. To use his own words, (from memory,) nothing can exceed the safety of these waters and their beauty. There is not a shoal in the Strait of Fuca, Admiralty Inlet, Puget Sound, or Hood's Canal, that can in any way impede the navigation of a seventy-four I venture to say, that no country in the world possesses waters such as these.

Olympia is a little town not laid down [on] the map by Mitchell, at the Southernmost portion of the Sound, on "Budd's Inlet"—named after one of the officers of the expedition which explored it. The Sound is a beautiful sheet of water, skirted all round by Bays and Inlets—all of which are very beautiful and picturesque. All, or almost all of these inlets have rivers emptying into them, making excellent seats for water power. Olympia is situated so as to command a very fine view of the bay, eight miles below. Chutte river comes in at one point, where another place is laid out, called New Market, in Simins' claim and on which he has erected a saw and grist mill. Somewhere on the Sound will rise a city that will exceed the New York of the Atlantic—a second Venice. *What* point it will be, *quien sabe*, but that it will be on the Sound and near the head of it, is as plain to one who looks at the map and sees what a scope of country it commands, as that two and two make four. Doubtless there will be numberless cities on the coast, but here will be the great city of Oregon—the head of navigation—commanding the region all below it as far as the Columbia, the Cowlits and Chickulees Vallies, and as far North as our possessions extend. The Willamette Valley, as I have said, is already thickly settled. Much no doubt yet remains unsettled, but when last year's emigration shall have got themselves homesteads the choice of lands, the water powers and mill seats, will have all been taken up.[79] Besides, a projected road[80] from Stillacom, (where there is a United States Fort,) to Fort Wallawalla, would divert the trail from the Columbia and Willamette Valleys. The emigrants, wearied with coming the 2000 miles, dread the additional journey of 150, that I have made, and hitherto settled there. Next year it will not be the case. I have no doubt the majority of that emigration will push at once, on arrival, for Puget Sound, as will no doubt a portion of this year's, and many of the old settlers, so that the settlement of the country and the increase of population will be unparaleleled.

Public and political attention is being attracted to it very much lately. I see one of the three routes proposed for the great Pacific Railroad is placed under the supervision of our new Governor,[81] who is to start by way of the South Pass from Minnesota only five or six leagues north, and crossing the Cascade Range, go direct to Puget Sound. If he crosses in the winter he'll have a "good time," though I fervently hope he may find a pass through. Of the three routes the Central one will be chosen as far the Pass, no doubt, but beyond "deponent sayeth not," 'cause he don't know. But it *will* be built.

To return to our own affairs. The country round about Puget Sound I have not seen yet. I have read, and been informed by those who have been over it, that it is a good country—prairie, with patches of timbered land. If big trees afford any evidence of a fertile soil, there are leagues and leagues of convincing acres covered with the tallest specimens I ever saw. The most common trees are the fir and cedar and varities [varieties] of pine, all of which grow to a tremendous size. Nine feet in diameter is not uncommon and height unlimited. I have seen a great variety of abuta[82] trees, evergreen, with red bark, fifty feet in height. The wood is very close and hard—excellent for furniture, but difficult to procure sufficiently straight. The yew tree is also indigenous here, though I have seen none as yet.

Twelve or fifteen miles from the Sound is a tremendous coal bed, which a company headed by Aspinwall[83]—the Pacific Company—are about to work. They have sent up engineers, who have surveyed and staked out a railroad, with a terminus to the Sound. This road will be eventually extended to Columbia river. The coal on the Cowlits cannot well be worked yet, as that river is not at all times navigable and *is* at all times dangerous, and the difficulty and danger that attends the entrance into the Columbia river is also a disadvantage. The Puget Sound coal has been tested and found satisfactorily to be coal. They sent up an "expert," who gave a report, shipped a small lot and experimented with it on a steamer at San Francisco. It answered the purpose, made steam, and gave satisfaction—the only objection being that in burning it made too intense a heat—a failing, I think, leaning (in this particular) "to virtue's side." It has long been know that there is coal at Vancouver's Island, but unfortunately, through the mismanagement of 54:40 men, it is British ground, and American capitalists prefer investing in their own country. The duty on it is 30 per cent, at least.

I have made mention in former letters of the great profits to be derived from the lumber business; a water-power and a mill seat is a fortune to a man. Lumber has never sold for less than $25 per thousand, and is now selling (and all that the three mills on the Sound can cut for six months is engaged) by speculators from California at $45. Two of them, after the Sacramento fire, came up Cowlits to the Sound, but stopped all night at Monticello, and meanwhile another and a sharper shap came along, and paying an Indian $50 (the ordinary fare was three or four dollars) to canoe him up to Wabass Port, beat the two—and then had the mortification to learn that another had "swept the field." The timber is emphatically the "tallest kind," grows down to the very edge of the banks, and only requires to be tumbled in. A cargo of spars has been consigned to China, and no doubt a regular trade will be opened with the Celestials. At present California is our market, and will be always.

Some seventy or eighty miles down the Sound is a beautiful Island, called "Whitney's Island,"[84] forty miles long, and from three to fifteen wide. The soil is very rich. It is called "the Garden of the Sound." There are but few settlers on it as yet, but many will go in the spring, and for a farmer it is probably one of the best locations on the Sound. At the end of it there is presented during the rainy season a very curious phenomenon. There are operations in the earth, which in the dry season are quiescent, but in rainy weather belch forth volumes of smoke, attended with a sulpherous smell.[85] The Indians are much afraid of the place, and give it a wide berth, when spearing salmon at night, or gathering oysters on the flats when the tide is down. In reference to these suggestive orifices, a farmer maybe would *rather* locate on the main land.

Let him locate where he will, however, and he has millions of acres yet before him where to choose, he is after all the most independent man in the land. There is no third man between him and his Maker; he sows his seed and the Almighty ripens it for him and I know of no place where the sower reapeth in such full and speedy measure. Peas are sown broad-cast, like wheat; potatoes planted, and nothing more wanted, till digged—then turned up with a plough. Plowing is done from September to July; wheat sown from October to May; potatoes in March, April or May. A team of two horses and a light plow can break up a prairie; a team of two yoke of oxen is better.

{I think as a rule the soil is thin, about Olympia I cannot say how deep it is, but twenty foot wells are still in alluvium. Really the climate seems to be of greater avail than the factor of very rich soil. Our florists have taught us that when they root in clear sand, having apparently no element of vegetable nutrition, plants are of vigorous and healthy growth. Here on ground not promising such results, are grown huge solid vegetables.}[86]

For a grazing country the prairies are unsurpassed. In winter (unless like unto the one we have just experienced—"may we never look on it's like again") horses, cattle or sheep require no shelter, and but partial feeding.

{The winters are mild, the blackberry vines in the woods do not lose their leaves. The great Japan Current sweeps in on the shore, and changes the whole condition of things, so that we, in the latitude of Quebec, west of the Cascade Mountains have a winter climate very similar to Savannah on the Atlantic.}[87]

The harvest in the Southern part of Oregon, is generally in the latter part of July, through August and part of September. The rainy season is the great drawback to one's notions of comfort, though absolutely necessary for the soil. There is but little rain from July to the middle of October, but "when it rains it *rains*."

{This is a wonderful berry country, if there is an opening anywhere in the woods, that place is immediately overgrown with a dense thicket of blackberry and other berry vines, that yield enormously both in size and quantity. The vines will be fifteen to twenty feet long.}[88]

I intend when I locate my claim (and I have one already in my eye) to raise at least vegetables enough to support me, and for this purpose want you to send me some works on gardening, farming, planting fruit trees, fisheries, cultivation of oysters, &c. Something *practical* and common sensesical, and also an assortment of seeds of the *best kind.* I send you some seeds of a great variety of berries indigenous to the country; the thimble berry, salmon bilberry, and sall sall berry.[89] They all bear fruit and are beautiful garden ornaments as shrubs. The latter is an evergreen with a rich velvet dark leaf. I enclose also a leaf of a plant[90] the name of which I do not know, but send it on account of its perfume, which is said to become more and more fragrant every year. This is but a part of one; when whole it is like a feather—fleecy and purely white, and as large as a May apple leaf. Ere you receive it, it may have lost some of its fragrance, for it was wet with the sea water as we came down in the canoe. By-the-bye—coming down that time we landed on the beach and gathered quite a quantity of liquorice root,[91] which grows here profusely. There is also a beautiful scented grass growing here—in appearance like ordinary grass, but exceedingly sweet smelling. Honeysuckles, different from ours, and giving out more fragrance, abound, with wild roses of every variety. You that are fond of flowers—and who is not?—would be more than gratified by the profusion of all kinds, many familiar to us, and more "strange and new," that in the spring gem the prairies and deck the hills.

"Springing in valleys green and low,
And on the mountains high,
And in the silent wilderness,
Where no man passes by."[92]

The Lemingia, yellow Ranunculus, a species of Trilliam, and Lapines,[93] are very abundant. I wish we had a branch of Adams' Express[94] here, (we *will* next year,) that I could send you larger packages of these, together with seedlings of the many esculent plants growing wild here, good alike for "man and beast;" as also some Indian curiosities, of which I have accumlated a great many.

Last season an expedition was originated here to go to Queen Charlotte's Island—which you will find laid down on the Northern Pacific coast—to rescue the crew of the American ship[95] you have doubtless read of as having been lost there and attacked by Indians. One of the party was my kind neighbor Billings,[96] who gave me an account of it. They found there a most singular race of Indians—light-skinned, blue-eyed, and red-haired, and very warlike—so much so, indeed, that prudence dictated that the rescuing party referred to should ransom the prisoners instead of demanding them unconditionally. This people seemed to have more refined notions of things, and a quicker adaptation to the manners of civilized life, (I don't mean in attacking ships and making slaves of their captives,) than any they had met. They had more comfortable houses and a less dubious way of living than our Oregon Indians, who depend principally on the salmon fisheries, and in spring time actually eat clover like Nebuchadnezzar, and "wax fat" on it. They were most ingenious carvers, too—their lodges being full of figures and imitations of animals, cut out of a species of porous slate.[97] Among others, (and which are the specimens I possess, and would like to send, had I the means,) are carved heads, full length figures, models of canoes, and a representation of the ship and the crew—the heads on these, *being white*, carved out of a kind of bone. They had taken carved "dogerotypes" of the captain and crew of the ship, which they sold to the party from whom I obtained them.

{When the ancestry of these people is traced, and also that of the Alaska indians, due consideration must be given to the wrecking of Chinese, and Japanese Junks upon these shores, as indicated by the facial characteristics of the tribes, and their facility in carving, in manner much like the Orientals, and unlike any other indians. Also in the square jaw of some of the Alaska tribes, and their hair of different colors, may be surmised the descendants of Prince Madoc of Wales.[98]

{In listening to the recital of some of the dialects of the north coast indians, by Geo. Gibbs, probably the best philoligist in indian dialects that the country has produced, something in the euphony recalled a boyish memory, and I repeated a garbled Welsh sentence I had picked up at school, from a playmate; it caught Gibb's attention at once and he asked me to repeat it, and when I told him it was Welsh he said it had a general resemblance to the dialect he was repeating. I suggested it was worth while to give consideration to the old Welsh tradition, and to anylize with that in view, it being my impression that those northwestern tribes had been greatly affected by not only the Oriental people, but possibly also by the Welsh.}[99]

Christmas

My letter is developing itself to a fearful, and to you, I am afraid, a most tiresome length. I must draw to a conclusion, and condense the remainder of what I have to say. And yet, writing at my window, and at times *revere*ntially, pausing with pen uplifted, as I look out upon this broad panorama of water, wood, and waving prairie, bounded by the distant coast range, which shuts us in from the great world, it seems to me I could gaze upon its varied sublimity and beauty, and try spasmodically to give you a realizing sense of it—forever. I intend resting here till the winter's severity is over, resting my ancle, myself, and the cattle, which, if I am fortunate enough to retain, and have them in order by spring, will sell for $200 a yoke, or hire for $10 a day, which will enable me to live in a tolerable good style, and afford me time and means wherewith to make explorations in the country interior, and all around the Sound. But of future plans more anon. We are living now, in comparison to our first four or five months' experience, in a tolerably "fast" style, and I am diligently following Romeo's advice to the lean apothecary, to "get myself in flesh," with a success that bids fair to give me a rotundency equal to that of a New York Alderman. There is plenty of material around us to conduce to this result; the shores of almost all the bays and inlets, or rather flats, are thronged with oysters and clams; the oysters small, but very good. The Indians who live about Olympia bring in a good many from the mud flat in front of the town, left by the receding tide, (which rise and flow here from 12 to 18 feet!) and exchange for other provisions, flour, vegetables, &c. But clams! we have "cords" of them almost at our own doors. Last night we went down on the beach below us, when the tide was out, and dug up a couple of bushels. From the door of our cabin we can shoot any quantity of ducks, many of which have already fallen victims to our voraciousness. For ten days I have been pretty much "confined to the house," in consequence of the continual snowing all that time. Such a snow was never before seen in Oregon; it was three feet deep. It is now, (Monday, 28th December,) thawing, and will soon disappear. It had made me somewhat anxious for the safety of my cattle, on whose existence so much depends. I went out on the prairie and found but three. These I drove among the hazel bushes; the others, I hope, are safe somewhere in the cops, where their own instinct would drive them.

Christmas last was the second I had ever passed from home. Only a year before I was up the Allegheny[100] river, and came "tearing down," as fast as the speed of the little steamer "Wave"[101] would let me, to be home in time for New Year. This Christmas eve found me again with you all in spirit—writing what was most within me, in this rambling incoherent letter. Christmas morning I got into my little canoe, paddled up to Olympia, and went out again on the prairie to see after my "floating capital"—the oxen. Before starting I borrowed Wilson's watch, for well I remembered our old custom of pledging each other on that day, wheresoever we all might be scattered abroad over the land, and precisely at the hour (allowing for difference of latitude) that you at home,—the "dwellers on the hills," and the brother far South,—were similarly occupied, as I have no doubt, I drank the silent wish—my vintage being purer than yours mayhap, for I was near no stream—but the toast was not less fervently expressed, though pledged in the fleecy snow. Christmas night, having found all my cattle safe, I spent a pleasant hour or two in Olympia, seated in the "sanctum" of the editors of the *Columbian*[102]—ate supper with Wilson, and then, as the tide came on, slid my little canoe down the shelving beach into the water, and paddled slowly down the Bay—looking far down, beyond "Gull Harbor," till the view was bounded by "Poe's Point," a broad silver lake, placid in the moonlight shimmering down, from the surface of which my paddle threw up the phosphorent spray like a snow drift—my course from Olympia marked by a gleam of sparkling brightness.

{Settled in the cosy cabin, and cheered by the melody of Brook's violin, with a panful of clams roasting in the oven, and some pleasant talk, a pervasive warmth, and a subconsciousness of the snowy expanse outside, and the sound of the crackling wood fire, life seemed pretty full, though tinctured with a faint drowsiness, to be lost

"In the world of sleep
The beautiful old world.
The dreamy Paradise of pilgrim thought
The Lotus garden where the soul may lie
Lost in Elysium while the music moan
Of some unearthly river faintly caught
Seems like the whispering of angels, blown
Upon Aeolian harpstrings, and we change
Into a seeming nothing, that is not."}[103]

The labor of writing this letter has been to me "a labor of love," and my confinement to the house in consequence of the continued snowing has enabled me to extend it to such a length—not to your weariness I hope—but I *must* close now.

"—My letter is almost closed—my theme
Has died into an echo—it is fit
The end *should* come of this protracted journey,
The torch should be extinguished, which hath lit
My midnight lamp—and what is writ is writ—
Would it were worthier!"[104]

But if it gave you an idea of the "perils by flood and field," the phazes and changes occurring to me since I left the aegis of home, or afforded any evidence, in its prolixity, of my wish to gratify your fullest anxiety, then has it realized all it was written for; while to me it has been "its own exceeding great reward," as bringing you all near to me, during the time occupied, than at any other time. And so, good bye. I could wish you were with me, this calm December night, to see with me the scene from my window. The placid waters of the Bay lie extended before me, undisturbed, save by the silent motions of an Indian spearing salmon, whom I can just see under the shadow of the tall pines; while fifty miles away, the declining sun (invisible to us,) gilds the snow-capped mountains of the Coast Range, from which Mount Olympus rises grandly to the clouds. I can imagine you are gazing on it with me, but as I add the last, closing sentence to my letter, I rise with a sigh, that "its but a dream" and that I stand ALONE IN OREGON.

At Home in Olympia

The following three letters (February 20, March 27, and May 22, 1853) were found in the Yale Scrapbook. Allen sent the first and last letters to his brother William, and the middle letter directly to the editors of the *Pittsburg Dispatch*. These letters introduced many new characters, and recounted Allen's interaction with local Indians and his attendance at a spiritual séance, among much else; they began with an entertaining account of his establishing a donation land claim and building a cabin on Budd Inlet, three miles from Olympia. The letters are interspersed with a number of passages from his Pittsburgh Manuscript, identified with braces { }.

Acquiring a Land Claim and Building a House

ALLENS CLAIM, (3 miles below Olympia,)[1] W.T.
Sunday, February 20, 1853.

I am under the roof of "my ain wee housie," now—and from within indite[2] you this letter—upon my own land, too; and with a strange proud feeling I look out upon the giant trees whose heavy shadows are grouped upon the beach, and listen with a freehold air to the music of the waves, as they curl in upon the shore, heralding the coming of the tide. My little cabin can scarce be seen from the bay, upon whose margin it is built—for the trees cluster round it so closely, growing denser and more dense back of it, until it seems almost as though the night never left it. At the right is my spring, down a little hollow, of beautiful clear water as ever rilled from a Pennsylvania hillside.

I have given you, in my former letter, some account of the Sound and Budd's Inlet. I am now on the opposite side from where I was then. I command the same extensive view far up and down the bay; at the head of the Inlet, Olympia is seen; across the bay is Quincy A. Brooks' claim—and far, far back of that rises Mt. Rainer [Rainier], a most beautiful snow-capped mountain, only to be seen in clear, calm weather, and, when seen, seeming but a few miles distant. It is a beautifully situated and valuable claim, rich land and heavily timbered, fronts upon the main bay or Budd's Inlet, at the head of which (you will recollect) Olympia is situated—and only half a mile across is bounded again by Ell's Inlet[3] or Mud Bay, upon which are procured the finest oysters upon the Sound; in fact it is upon a kind of peninsula, bounded by those two sheets of water, and, being only half a mile across, just makes a quarter section—half a mile square it is about four miles long, and is now all occupied.

I will now tell you, in my journal kind of way, how I came by it. I wrote you last, when a guest at Brooks', where I was having a glorious rest, as were also my cattle. When the snows had disappeared, and I began to feel strong, I went out with the "wood-choppers" on the place, and tried a little experiment to see "what was in me," found I could, at that, earn readily $3 a day, and give it a good test by continuing at it ten days. Meanwhile I was looking round, keeping my eyes open, with ulterior views which I shall not expatiate upon now. I found the claim I had in progress of purchase would take some time to bring to a conclusion, and meanwhile there were equally good sites lying around, to be had "for the asking," and a residence on it (of yourself or agent) four years.[4] I thought, while ulterior plans were in abeyance, I might as well secure a homestead to radiate from in this way. I had become very much attached to one of the young fellows I had chopped with, originating in that peculiar feeling of reciprocity incident to chopping at the same tree, which an after intercourse had ripened into friendship. Shirley Ensign[5] was his name—a genuine whole-souled fellow, from Ohio. I took a little trip of exploration, (starting on Sunday—I am sorry to confess, thus making "a Sabbath unto yourselves,") found this locality, was pleased with it; returned and together we went over to look at it, and (feeling more and more pleased) went over one sunshiny morning marked out the locality of our house, sung "Hail Columbia," and stood on our own ground, "landed proprietors, be jabers!" There were three of us—Shirley, myself, and a Dutchman[6]—a pretty intelligent fellow, and a "Shukum"[7] chap with an axe—and we took that peninsula instanter—would have "annexed" a continent, had there been any one "to the fore," and the stringency of the land law permitted. Though in principle we were all Free Soilers,[8] we had another strong-armed fellow[9] with us, whom we hired at three dollars a day, (to chop wood, and make himself "generally useful" between whiles,) boarding included, in consideration of which his wages were not to commence till we were all under the shelter of our own house. On Monday morning, bright and early, two of us went up to Olympia to purchase some lumber and "much-a-much"[10] provisions, the other two remaining to frame the house, or rather clean a space to build it on.

Lumber is now hard to get, from it all having been engaged (as I write you,) by speculators from California, to supply the demand occasioned by the annual fire they have at San Francisco. I was fortunate enough to get a thousand feet from friend Simmins[11] at the minimum price, $25, and paid five dollars more to a couple of "Sowahas" (Indians) to take it down, together with a rousing cooking-stove for $40 more, some bread, potatoes, and meat. On our return, the same evening, we found the boys had made a little start at the house, and we all merrily turned in to help them, and tall fun it was, if hard work can be considered a joke.

The first evening we built a glorious big fire of the logs we had cut down, and two of us, (the other two crossing over to Brooks' for a shelter,) laid down to sleep. We slept soundly, too—after watching the Indians spearing salmon by torchlight, down on the Bay. I was awakened from my slumbers by a loud "ugh!" and starting up, found an Indian standing over me. It startled me somewhat, as, in the indistinct light, I did not know but what it might be the panther we heard screaming in the wood back of us when we laid down. The Indian was not less frightened at the sudden dash we made for the axes, laying at our side, but he soon recovered from his surprise, and we from our fright—when he exhibited a large string of fish he had come to sell to the "Boston man" as they call Americans, in contradistinction to the term "King George's men" which they apply to the English.

It was a week before we finished our house; during that time we worked all day, (and none harder than our "man servant," who was anxious not only to be under shelter, but also to have his wages commence,) and at night, by a reviving fire, slept under no other roof than "the broad cannister of Heaven," as Mrs. Partington[12] calls it. Luckily, during all this time it did not rain, the supply for that year having been already apportioned by the Power that supervises those little arrangements, and so our sleeps were glorious and refreshing. On Saturday night, when you at home were probably sitting around the fire blessing the darkness that ushered in the Sabbath rest, we four entered *our* now newly completed home, and building up a blazing fire in our stove, were grouped around enjoying its grateful warmth—engaged in various "household affairs," and at intervals talking of our plans and future prospects, interspersed with many a furtive allusion to the homes we had each of us left "afar off." The night came down on us with its deep shadows, and through the open door we could see Mt. Rainer's white top grow dimmer and more indistinct—and still we talked and "all our theme was friends,"

"And all our talk was frank and free
As talk in the forest's heart *should* be—
Thoughts of different years and moods were we!"[13]

It is at such times the heart is busiest with memories, and those that have never known what it is to be isolated from the amenities and comforts of home, can but little appreciate the recurrence of so many allusions to their loss. Were any eye but yours to peruse these letters, the frequency of them might weary, mayhap—but an "Oregonian" well knows they "well up" from the heart. The feeling seemed general, and extended even to the "strong man" who had never been from home before, and who narrated with a comical gusto the performances of a favorite "boss" at home, and expatiated on the charms of a "Miss Nancy" left behind, with a pathos that was infectious. I, in my capacity of "cook and bottle washer" generally, was kneading viciously some dough I was preparing cakes from for Sunday's breakfast, and striving to repress the choking sensations excited by Mumpford's reminiscences, when Ensign (who had'nt said a word all night,) from out a corner struck up in a clear baritone voice the beautiful air of "Dundee." Mortal man could'nt stand that; and, signing to him to quit instantly, I dug my fists into the dough with an energy that threatened to break the pine bottom out—and succeeded, by most wonderful self-command, in preventing my "feelinks" from proving "too amny for me."

Next day was Sunday—and, though

"—the morning breeze was bringing,
No solemn sounds of church bells ringing,"[14]

to our ears, there was none of that party I believe that was insensible to its calm quiet, or unmindful of the duties and frame of mind its institutions imposed. True, we had no "fretted arch" to set under, or "haspit pew" to set in—but around us were the tall trees, whispering His name "whose Temple is all space," and "a blue sky over all," and so calmly and quietly (in our own house, or out on the margin of the bay,) we sat and talked of home, (which is the nearest approach to a sabbatical mood I know of,)—and so the day waned away peacefully.

"Sabbath holy!
To the lowly
Still thou art a welcome day;
When thou comest, earth and ocean,
Shade and brightness, rest and motion,
Help the poor man's heart to pray."[15]

{It may seem a wild stretch of fancy to think of these old forests as having known so many centuries, but trees have been felled, whose rings of what we accept as yearly growth, have numbered over eighteen hundred.}[16]

Progress is Made

Monday morning to our work again, buoyant and exultant, each and all to our several departments; for more facility in my own I put up some shelving, and rigged a little table in a corner near the stove. Ensign cut down a huge fir, obscuring the view in front, which took him a long while, as it was a tough old fellow, probably four feet in diameter and over two hundred and fifty feet high.

We were compelled, indeed, to cut down many of the trees near our house—not from that Vandalism that impels a snob to cut down all the old forest trees, and then immediately proceed to plant new ones, but because those in the vicinity were so high and slender we feared (when old Boreas would be about) they would snap off and "mameloose"[17] some of us.

Fred was meanwhile busily engaged in making and hanging a door, the latch string of which was left outside, for no man locks his house in this country. My cooking facilities being completed, I made some window frames and put in some sash I had bought at Olympia; and, although *we had no glass*, the house looked somewhat finished. We had bought plates, &c, and *capacious* spoons—for we were determined to live in a style superior to any in that "neck of country." {I had taken lessons from an expert in Olympia in the matter of Clam Chowder and could make a compound that would make any man forget his yearning for home.}[18] Our beef *cost* us sixteen cents per pound, potatoes $2.50 per bushel, turnips 12 ½ cents each—but they were "whalers," I assure you. Our barrel of flour cost us $40, a barrel of salmon, (the cheapest thing in the country,) $15. These, with sugar, coffee, tea, &c., enabled us to make quite a respectable provision for the future. The house, when done, was a primitive kind of an establishment—being formed of rough boards, set on end, but a *shingled roof*, on which we prided ourselves mightily[19]—as well, indeed, as on the whole finish of the house, as there was none superior to it around there, except in Olympia, where they have arrived at the dignity of clapboards. The whole concern, including a canoe, and provisions sufficient to last us a month, cost $210; probably at home would only have cost $70.

Next morning we shouldered our axes and went out to work; there were five of us now, for I had that morning hired the faithful young fellow who had brought my cattle on from the "Cascades," on "my own hook," he having come all the way up to the Sound, and wanting to join fortunes to mine.[20] The result of the first week's operations are as follows: We have cleared a space around our house to plant a garden, and hewn out 1,200 feet of square timber, worth in cash 12 cents per foot, or 14 cents—which it has hitherto always sold at. So, after paying all expenses, I think we have made "tolable" good wages—certainly more than I used to have allowed me in past times, when in the "grocery connection."

There are three claims here—I think four—we have not as yet had time to measure; we hold three, at any rate, and will try to hold four. Dr. Eggers,[21] who came out with Brooks,[22] will take the remainder. We were, I think, very fortunate in securing these claims; as the emigration, as I have before intimated, will be heavy from the Willamette and Columbia valleys, and most all of next Spring's emigration (we hear) will push directly, without stopping, to this place, and all "Sound" claims will be valuable. So I am losing no time, nor yet shutting myself out from other prospective views, one of which, (and the foremost) is to secure for William (by purchase) a claim already settled, and the title of which can be obtained as soon as it is surveyed.[23] Fitting it is I should, for our interests as well as affections have ever been as one, and we

"Are now but *two*,
The other sleeps in Death's untroubled night.
We are now but two! and we will keep
The link that binds us bright!"[24]

The other project—but no matter; it is not my way of talking about a thing till I have accomplished it, or at least made such a commencement that I can safely say it *will* be realized. I have alluded incidentally to the wonderful advantages Northern Oregon or "Washington Territory" possesses over Southern Oregon, in the wonderful faculties of ocean communication at all seasons, as well as an equally good soil on the average, equable in climate and development of mineral wealth, even including gold—which, by all purity of mineralogical reasoning, must be found here, and which the recent discoveries at the Dalles confirm. This, however, is a future advantage that will be developed in "the fullness of time." The present superiority consists in the other advantages enumerated, and particularly the extensive lumber trade, which will be a mine of wealth to the country. The southernmost part of Oregon, like California, will always have to depend upon us for its supply. Timber in both countries is sparse, except some districts far from means of transportation. The Columbia cannot be entered at all times, and the people of San Francisco are in the semi-annual habit of amusing themselves with a destructive fire to break the monotony of their calm and quiescent lives.

Ships at all times can enter through the "Straits," and come far down in the heart of this lumber country, anchor close to the shore and have the three hundred feeters toppled down on their decks, if they wish to save labor. While I am writing this *there are three vessels anchored in sight*, from California, loading with hewn timber, compelled to take it there to have it sawed, for the only two mills on the Sound have more than enough to do. One of these mills (owned by my friend Summins [Simmons], at the mouth of Shute river,) is a water mill, running two saws, and costing (when labor was high) $5,000, and rents for $800 a month! The difficulty of getting machinery out here, (fast decreasing now,) and more than all the immense distance making a residence here necessary to supervise it, as well as the country being so little known and its resources as little developed, have hitherto prevented capitalists from erecting steam saw mills, but a year will work a great change, and the first erected will make a fortune—15 or 20 would not supply the demand. I know of nothing so profitable at present as embarking in this business. If my strength and awkwardness, and the assistance of a devoted brother,[25] will effect it, I'll have one on "Allen's claim" that will rip out 1200 feet a day, or "perish in the attempt!"

It is a strange kind of a life a year's change has inducted me into, and yet withal a pleasant one. In consequence of a superior skill in cooking, I am (as I have before stated) "chief cook" to the establishment, and pride myself somewhat on my catering. My culinary duties engross all my time and undivided attention, all morning. In the afternoon, I join the rest with my axe, and, first "*ax*ing" them how they are, proceed to work slashing into the trees around with a "vim"—carefully venting a grunt at each stroke, to aid the descent of the axe, in true orthodox woodman's style. It is a healthy manly employment, and I can carve my way into a pine tree equal to any one on the ground.

Woodsmanship

{Just above us is a pretty little bight,[26] or bay, where we are getting out our timber, to walk from our cabin takes us through the brush and lengthens the distance, so after breakfast, the men take a larger canoe we found it necessary to purchase, and paddle across the bay to their work where I soon hear the ring of their axes, and their cheerful voices. Maybe a few verses of "The House Car-pen-ter" floats over the water.

{I remain and put everything in train for dinner, and clean up the dishes; we are cleanly here in a crude way and do not follow a common custom in the bachelor cabins, of exhausting all the reserves of one side of the plate and utilizing the other, before any washing is done.

{This completed, I take my axe and in the smaller canoe also paddle over and take my place on the small end of the tree, and score for the broad axe to follow.

{To be a good axeman requires a strong wrist, and a good grip, as I have neither, I have to make good by taking that part where the lack of these factors are less felt. In the squaring of a great tree trunk, at the heavy end, there will have to be great "juggles" taken off, say eighteen inches thick, and six to eight feet long, and this requires strength and dexterity; the juggle is cut in at either end to the depth that would square the log, and an axeman at each end acting in concert, split the juggle off; the one makes a vigorous blow that maybe sinks his axe in past the point of ready extrication, the other strikes an equally strong blow in precisely the same line, that opens the crack and liberates the others axe, and so on alternately, moving along the juggle, until it is split off, and falls to the ground, when the next juggle is taken off in the same way, the juggle being thicker or thinner as the movement is toward the top, or butt, of the tree; at the extreme top, if the log is small, say ten by twelve inches, there may be no necessity for any juggles, as deep scoring may reach into the squared line, so that the broadaxe may cut through to the mark and here is where the weaker axeman may, with quickness, and three strokes to the two, of the skilful axeman make amends for his lesser strength. And here is where our humble servant justified his right to live, and demonstrated his equal usefulness.

{Do I seem to take too much space for this detail, that would not much engross the reader to whom the art of squaring timber has no interest? Let me venture to say, that it is no light thing when you find yourself in a position where your physical condition and previous habits of life unfit you to rank among your fellows, to make your brains make amends for your lack of muscle, and places you on a equality that enables you to maintain your self respect.

{To be an expert woodsman is not simply a matter of muscle, he learns to know the woods, and though generally his interest is not that of the naturalist, but is confined to the necessities of his calling, he absorbs the spirit of the forest, and loves it, and therein we have companionship, and he has much information to impart, that haply you may make other uses of, and benefit your soul.

{At eleven, I leap down from my poise upon the fallen tree, and going through the brush to the beach, get into my "one man canoe" and paddle back to the cabin, and get the dinner, now and again looking through the window to the stately Cascade range, crowned by lordly Tacoma,[27] and the dinner getting rises out of its tedious detail, and becomes a function, with the mountain as a looker on.

{My lungs have grown capacious, and can hold great and prolonged sounds, and when I stand upon the stoop and send forth the continuous call that is known as "the Allen whoop," I know it is resounding in the forest where the men are working, and the bright axes are struck into the tree and with an answering call they troop down to their canoe and paddle over the bay, to where their dinner waits them.

{It might under other conditions be a longer passage, but their paddles are plied vigorously, and they are soon coming up the little ascent to the cabin. A hearty meal in which lies no after "heartburn," the after pipe, which brings ease of mind, and a generous opinion of all mundane things, and we are all off together to the woods again.

{And so until five, when I again take my canoe, and at the cabin prepare supper, and when I make my call, hear the same joyous response, and have the same quick return. Maybe it is salt salmon, and potatoes, with the excess salt eliminated by a nights soaking in the running stream, or clam chowder, where clams have not been forgotten in the mixture, or maybe roasted Mallards, or fresh fish, which however is generally a breakfast dish. I am filled with admiration for the brilliant "Chef" who discovered the fitting dishes for the different meals, but I give no meed[28] of praise to the adjuster of fish, fried potatoes, or eggs for the matutinal[29] meal. That was instinctive, the first fire[-] using barbarian would have made that distribution of them had they been at hand.

{After supper, came a quiet smoke, out on the stoop, or if it seemed too civilised, some branches of cedar and a full length upon the ground, with maybe a little curl of grey vapor from a slight fire, because as it stole upward, it gave human influence to the stilness.

{Later such books and papers as we might be the fortunate possessors of, and I to my home letters, which had no limit as to subject or time, and kept me intent long after all the others were deep in sleep.}[30]

Indian Neighbors

We rise before daylight in the morning, and while the day is breaking have a glorious view of Mt. Ranier, immediately behind which the sun rises, throwing it out in grand relief; first, in the haze, seeming a dense heavy mass, gradually growing more and more distinct, until it assumes a faint tinge of crimson, then shines all goldenly until the sun rises from behind it, when it stands out a pure white peak—dazzlingly bright and "beautiful exceedingly." When night "draws her crimson curtain round," we can see the Indians fish immediately below us, the light from their torches dancing on the water. They sometimes build a fire in their canoes, of pine knots, and gliding silently over the shoal water near shore, they see the fish (attracted by the light) in six or seven feet water, and spear them with the greatest dexterity. Their spear is a long thin rod of cedar, three-fourths of an inch in diameter, and about ten feet long, furnished at the end with a piece of bone, backed like a fish hook, with which they strike the fish and prevent their escape. They are remarkably expert, and rarely ever miss their aim. Sometimes they fish at mid-day and are equally successful. They are also good shots, bring in a great many ducks, and exchanging them with us for our "store victuals," which to them are a luxury.—They bring us several varieties of fish, and all excellent eating; a species of Perch and Bass, both of fine flavor; Flounders I repudiate. They also have a Sturgeon, but I have never been able to purchase any of them yet.

One old chap comes regularly nearly every day; he has just been here, after a two day unwonted absence, has been sick, wanted some bread, will bring some fish another time; he was very hungry and I gave him some; he returned in an hour after with a fine string of them, on a wythe, and flinging them on the floor, they lay floundering about appealing to me as cook to come up to time and go into the scaling performance, "Wake Klosh."[31] These Indians are a singular people. They lie around until hunger forces them out, and then shoot or fish to satisfy their cravings as hunger demands.

{Providence does so much for them, with the waters full of all kinds of food at all seasons, woods teeming with succulent berries, and a climate that is the acme of luxury without enervation, with that saving exception, that emphasises all the rest, the winter rains, which are kindly in their way, and ensure to us all the summer's unblemished days.

{Providence indeed does so much and their wants are so simple and so fully within the limits of such provision, that their own endeavor has become inert, and their provision for possible contingencies was at all times negligable, and with those who have come in close contact with white men, it has grown less, that it was not unusual for others than my old chap to be suddenly destitute, with but slight stress. I think they were more thoughtful of provision for the future before white men came amongst them. In the tropics where all seasons are fruitful ones, and garments were only worn as a concession to prudishness, the aborigine might leave his whole future in the hands of his gods, but even a few fruitless months annually, would have compelled some slight "sweat of his face."

{Maybe we are over critical with Lo, I seem on reflection, to remember large classes of men at home that at times lapsed suddenly from even extravagance of living to sudden poverty, from temporary cessation of work, when it would seem the prosperous days should have given accumulation to have met the emergency. It was a wise hen which said to the horse in the stable "let us agree not to tread on each other."}[32]

Indian Legends

They are very superstitious, and have many traditions concerning many parts of the country, many of them not devoid of romance and poetry.

On the Columbia, near the Cascades, there are numberless falls; it is a very singular weird looking place altogether, and their tale concerning it seems not at all improbable in some of its parts. They say that Mt. Hood and Mt. St. Helens were once very great friends and they had a bridge built between them, that one might communicate with the other, but a little "miff" breaking out between them, (originating in an angry dispute as to their respective heights,) they threw the bridge down, and have not since "spoke" each other on any topic. It certainly looks much as if at one time there had been a natural bridge there, and the swollen waters of the Columbia, and the volcanic changes that land has undergone in the "far back abysm of time," had broken it down; great masses of loose rock lie down upon the cascades, at which point the river is very narrow and the cliff on either side very high.[33]

They are all imbued with a great dread of Mount Hood, and cannot by any means be induced to accompany any one making the ascent. East of us, near a mountain, they also tell us of a lake, where the "old gentleman" lives—and, from their accounts, has a pretty jolly time of it, occasionally showing himself in the form of a huge serpent. Deer driven into this lake, sink at once, they say, and nothing can possbly live in it except the aforsaid serpent. Somewhere in the same locality they tell of a prairie, high up amid the mountains, and surrounded by hills, which the moment an Indian places his foot upon it opens and receiving him into it, closes again, in its unwelcome embrace, to a living death.

{Of a little lake near Olympia, they tell a more possible story; it is situated a mile or two from the bay, but they tell of an indian being drowned in it, and his body some time afterward being found in the Bay. The prairies have many beautiful clear and deep lakes with fish in them, but having no visible inlet or outlet.

{It is not impossible that the drowned indian may have found the outlet, and subteraneously made his way to the sea; it is possible one indian was so drowned in this lake, but it is certainly no other will ever be, for you could not drag one into drowning distance of it; I have seen this lake, it is not large, is surrounded by deep woods, is thirty feet below the surface, has great fallen trees round its shores, and is a dismal evil looking place.[34] The Loons, wary as they are, keep open house there; as I passed, one uttered its prolonged manaical shriek of laughter and was answered by its companion.

{In quick impulse I hurried on, and a little shamed, turned back and took a longer look at it. I believe I have unusual command of my nerves, but as I looked, I wondered, if by any chance I should fall in, I could swim through those tangled branches to a landing, without a quickened pulse, and singularly, I really thought it might be doubtful. In the indian tongue, it is called Devil's Lake, and the name is all right.}[35]

There is always something upon which these traditions are based, and an examination of these localities will no doubt some day disclose something extraordinary.

The Indians' Fate

These Indians are fast meeting the inevitable fate that seems to be their destiny elsewhere; their numbers diminishing here with a rapidity as fearful as the small-pox scourge among the "Black feet."—fast fading away, and soon their history will be "as a tale that is told;"

"By the rivers lonely marge,
Rotting is the Indian's barge—
And his hut is ruined now
On the Rocky Mountain's brow."[36]

Many causes contribute to this fatal result. Their shiftless manner of living, their unnatural mode of promiscuously marrying among each other,—sisters and cousins, and I expect they are not over scrupulous about their maternal relation—so that, within the memory of the settlers here, their race has decreased four-fifths! They themselves are well aware of it, and for a time hailed the approach of the white man (to whose arrival their traditions allude,) as their deliverance—and this will account for the cordiality extended to the first missionaries who came among them, the most war-like of their tribe being friendly. I heard an eye-witness relate a touching speech made by "Bloody Chief," one of their noted warriors, then in his old age—at a meeting of the missionaries and two or three of the hostile tribes, in the beautiful "Grand Round" valley, once the "dark and bloody ground" of their wars:

"I speak to-day, for to-morrow I die. I am the oldest chief of my tribe, was the high chief—when your great brothers visited this country. They honored me with their friendship and counsel. I showed them my numerous wounds, got in bloody battle with the Snakes. They told me it was *not good*—better to be at peace; gave me a flag of truce. I held it up high overhead, talked often after with my tribe in council—but *never fought more*! Clarke pointed to this day—we have long waited. I sent three of my sons to Red River, to school, to prepare. Two sleep with their fathers. The other is here—and can be ears and mouth for me, for I can say no more. I quickly tire. My voice and limbs soon tremble. I am glad to meet you, but shall soon be still and quiet!"

Poor creatures! Following close upon the footsteps of the missionaries, come the advance guard of that horde of Saxons,

"In multitudes like which the populous North
Poured never from her frozen loins—to cross
Rhine or the Danube;"[37]

that is destined in time to overrun the whole continent, and year by year they were compelled to see their lands occupied, their fisheries appropriated, and the game made scarce and more scarce, till in their grief they were ready to say,

"Ye waste us—aye, like the desert sun
In the warm noon, we shrink away—
And fast ye follow as we go; till we
Are driven in the Western sea."[38]

They are a friendly race, or long ago would they have resisted the encroachment of the settlers. *Their lands have never yet been bought from them*, and they are anxious to have a treaty ratified, securing them a "reservation" of some of their lands and security for future existence. They say the "Tice," or royal family are nearly all dead—and soon there will be no body to represent them. Mindful of their rights, I endeavored to preserve *my* conscience clear by making all right with them, so far as *my* claim was concerned. They live such a wandering Arab-like life—moving every year to some fishing grounds, the falls of some river, where they remain all season to lay in a sufficiency of salmon to subsist them during the winter, that it is difficult to tell who to purchase from or treat with. After complying with the usual customs and requirements of *our* laws, to secure possession of our claim, we looked around for the original owner, and found the old fellow who brought us the fish on the first night of our entering possession was the ostensible one—at least he could say "my right there is none to dispute," till *we* came—as he had been in the habit of making it his half year's residence, with some four or five others, for some time. Accordingly I proceeded to buy out his interest, and right and title, and to give it impress and dignity we dressed ourselves in our best, summoned a solemn council, selecting our treaty ground under the biggest tree in the neighborhood, and after a solemn silence of half an hour, succeeded by a big smoke (which part of the performance Mumpford and the "Siwashes" enjoyed most,) the conference opened in sonorous and gutteral "Chinook," eked out by expressive pantomine and energetic grunts, and after some time spent in stating the "object of our meeting," and arranging terms, a solemn treaty was concluded between us, the copy of which shall be sent to the archives of the "Oregon Historical Society," when we have one.

It would be superfluous in me to detail the conditions and "considerations" given for our little Peninsula, mayhap to one accustomed to the figures appended to the sale of a burying lot, even at home, he might be induced to think (like Penn's purchase of our own State,) it was a deuced good bargain—and the amount paid was like a rich man's contributions to the velvet bag, that forms part of our church worship—"nothing to nobody." It matters not, however; the Chief was satisfied—though he evidently thinks (from the frequency of his visits since to us) that he has acquired a reserved right to be our guest as long as he lives. I shall not contravene him—a plate shall always be placed for him so long as I retain my situation as cook. I would respectfully suggest that Government should humbly *imitate our example*, and *pay* these Indians for their lands, while they are yet in existence, and ere their chiefs have not all gone "from earth forever." The greater part of them live down near the mouth of the Sound, the "Hyas Tice" (high elders) have nearly all been removed by the "Hudson's Bay Company" during their occupancy of this country under the title of the "Pugets Sound

Agricultural Company," (who with a capital of two and a half millions, thus hoped to secure the best localities, after this part had passed out of their country's control.) "The Duke of York," "Clarence," "King George," and others—who have themselves with as much dignity (and more morality) than the Pohntostes[39] they were named after—the highest "Hyas Hyas-Tice" (very highest chief) of all is old Seattle,[40] who retains his original name. One of the embryo metropolises here is named after him—and in excellent taste, I think. The least we can do is to leave them their names, after depriving them of everything else.

When I was last down in Olympia, the numerous lodges of Indians there had turned out unanimously, and were busily employed in the avowed object of chasing the Devil out into the adjacent woods. Some among them had been sick, and their "Tamawanus"[41] or medicine man had cast his "brimstone majesty" out of the "possessed," and, for fear he should return, had "formed a circle," and were running him out of the vicinity. They have done this so frequently that the woods must by this time be populated by the sulphurous imps. But still the game goes on. Whether they believe in an infinity of devils, or that the same old chap possessed the attribute of ubiquity, or goes out and comes back, I do not know. Do *you?*

By the way, I may say I had the honor of introducing a new specimen to the country, in the shape of the "Old Knick'-erbocker Magazine"—a copy of which I found had preceded me to Portland, (sent by some friend,) and which, as "a treasure shrined and cherished," I brought all the way over the "Pack Trail" to Olympia. My friend, the genial editor thereof, may thank me for making him "in strange eyes not a stranger," and his Gossip in this far land, a gladdener of many a lonely hearth. Let him take note of it; for I believe *that* number to be

"The first
That ever burst
Over *this* silent sea."[42]

It, and two or more numbers, since arrived, are the principal attractions of our cabin, and, together with an old sun-bonnet, (which one of the boys fished up somewhere as a prize, and hung up in a conspicuous place,) constitute the "lares" of our home.

Giant Trees and Many Fish

I have, in former letters, alluded to the immense size of the trees growing here. "Oh! ye of little faith," who have heretofore doubted the strange stories "you read about," listen unto me, and I will, (seriously speaking) in reference to "what mine eyes have seen," speak the words of truth and soberness. You have heard of them being 300 feet high? The other day I took a notion to measure one *we* had cut down; not because it was by any means a large tree, for it was not as large as the adjacent ones, but it measured three feet in diameter, and 185 feet high. In coming up on the "Pack trail" from Vancouver, I saw many that I measured that were *ten* feet in diameter, which to the eye held out in proportionate height. This would make *them* 730 *feet high!*[43] This seems impossible when one look[s] at the figures, but every one I have asked say they *appear* to hold the same proportion as well as this. No one that I have ever heard express an opinion places the "twelve footers" at less than 400. I will, to please you, knock off 200 feet from my first estimate, and *call them* 500—won't say a foot less. I have seen some that measured *fifteen feet* in diameter; I leave you to cypher the height of these. Many of these rise 100 feet without a branch.

Talking with an old residenter the other day, concerning an article I had read in the New York *Tribune*, concerning the culture and raising of fish from the Milt—taken from a report to the French government, embodying the result of scientific experiment in the river Seiene [Seine]— he gave me some curious facts concerning Salmon that may be new, or not generally known. He says that up a stream where there are two forks, the Salmon of each kind will go up separate forks. There are generally but two kinds, in the main body, they are of course found in the same stream together, but at the forks each variety will take up an opposite branch. If you see a certain kind of Salmon in one fork you may be sure to find the same variety in all succeeding years—"curious, ain't it?"

{I was told this also at Warbassport on the Cowlitz. Where in the wide ocean these Salmon go after the spawning season, is not known, nor how they spend the two or three years before they are large enough to return, no one knows, but it is extraordinary that at the proper time they should, out of the immensity of the ocean, find their way back to the stream in which they were spawned, and even into the branch of it from whence they came. If the homing instinct is so strong in the cold blooded fish, what wonder that afar off the human being's thoughts turn yearningly towards his old fireside, with all its loving memories.}[44]

Spiritual Manifestations

We have had visits, lately, in Oregon, from reputed spirits.[45] For some time back we have had rumors of "manifestions" being current in Portland, and have been anxiously awaiting their coming here, but suppose the rainy season, severity of the weather, and the hardships of the "Pack trail," have hitherto prevented their arrival. They came at last, however, and have spelt out some curious observations, at Olympia. At the first meeting the spirit of a ship Captain, who had been lost overboard in his passage to San Francisco, was present, and told the exact time of the accident, although no one in Olympia, (where he was known,) knew the time. When the vessel returned, an examination of the logbook showed its truth. I have never been able to attend any of these meetings at Olympia, but twice I went on board a schooner that lay in the bay, the Captain of which was reported as being a medium, but though the Captain was a very clever fellow, he could procure us no manifestations. It was a curious sight, however, to see us all assembled there in his little cabin, crowding and grouped around the little table, all hushed, awed, and silent, in "anxious expectancy"—waiting for the faintest sound or sign. The men of the "fo' castle" were as silent as we and peered curiously down through the half open hatchway, wondering what our object was. The Steward, a shambling, slouchy fellow, with half a fathom of slack in the rear of his flannel breeches, shuffled around the stove on pretence of keeping up the fire, but soon disappeared on deck to tell the wondering sailors that "we wos a magnetizin'." We sat there on these two occasions, for over an hour, hearing no sound save our own breathing, and with no interruption, save our medium breaking the solemn stillness by an occasional order to the mate above concerning the "mizen," which I suppose had no connexion with spiritual matters. But "wake"[46] we stept into our frail canoe, (very cautiously for it was a frail concern, and very apt to "keelapie,"[47]) and steered for the "light shining in darkness," and in our own little house we gathered around our table and tried it again. But they never came and we went to bed, nor *have they ever come since.* In Portland they have tried it with success, and in San Francisco it has full swing and many believers.

There is something very extraordinary in the rapid march this belief, mania, delusion, truth, or whatever you may please to call it, has made since its origin—not more than two years ago, in the interior of New York—and yet in this space of time it has "girdled the earth." The philosophers of France are frantic with this new philosophy. The Germans are a hundred years ahead of any other people in their development of it.

The "Autocrat of the Russias" is said to have had his belief confirmed in its truth by a private assurance from Peter the Great, that he can "go it" in his schemes against Turkey without fear of "intervention"—old Radisky backs the advice. Even the Pope and his cardinals are said to have had private conferences in the Vatican with Luther, who repents of his life and labors. St. Peter is said to have assured them to be of good cheer, and the attack on the School System will yet be sustained. But I really cannot see the benefit of opening up an *extra* communication with celestial regions, if the Holy Father possesses the power his sincere believers attribute to him. The subject seems to be of general and absorbing interest. The greatest intellects and purest heads of other countries, as well as the best in our land, are absorbed in it, and German psychological works, the "Science of Prevoust," Jung Stillings' Pneumatology, Richenbach and Swedenbourg,[48] (the latter I am glad to see,) are sold and read with an avidity never known before.... The Paris Academy of Sciences has recently announced the result of an experiment (among whom, with many other "Savans," Humboldt was a witness of) that proves the control of mind over matter. *The direction of the compass being actually moved by the effort of the will.* This I should suppose would put beyond question the existence of a power we know as yet but little of. So we must wait and "bide our time." I suppose "my faith is large in Time and He who shapes it to the perfect end." I (confidentially speaking) cling to the "spiritual origin" of it. There is something inexpressibly subduing in the adoption of a belief that seems to have prevailed in all ages, and is instinctive in every one's heart—a belief that the "loved and lost," who have "gone up before," are ever more around us—not as *bad* spirits, helping us to do evil, but as "ministering angels" suggesting good counsel.

"It *is* a beautiful belief,
That ever round our heads
Are hovering, on noiseless wings,
The spirits of the dead."[49]

With such sentiments I can have but little sympathy for the many who *are* said to have been driven to madness by this belief. I fear no evil from spirits—rather, in view of the long prophesied millenium, and the comparatively little progress we have made to that state of purity, *necessarily* heralding the advent of that time—can convince us of no more efficient means of accomplishing this result under Providence, than in communications from those whom we loved here on earth, and would fain rejoin in the "Blessed Land." Have we really become so debased and impure since our common mother displayed that fatal fondness for fruit, that we shall never again be restored to those primal days of spiritual prayers, when the "angels come down on earth" and Abraham "entertained one unaware." I can conceive of men, even in *these* days of materialism, of such blamelessness of life and purity of heart, on this earth "whose spirit walks abreast of angels." Peter Neff was one of these, the Pastor of the Swiss Valley followed to his last home by 20,000 of his flock—Swedenburg, another, who after a cloud of hundreds of years resting on his name, is just merging into appreciation, and many others.—History is full of instances where

"—every now and then
There lived some pure soul to whom
They were visable again."[50]

Why not to all?—who would not welcome it?

Surely they *are* all about us,
'Tis our senses that are dim,
That we cannot see like Jacob
Cherubim and Seraphim—
That we cannot hear the eternal
Music of the Seraphs' Hymn."[51]

Something too much of this for an Oregon letter, mayhap, but—"these are my public sentiments *privately* expressed.

About the Cabin

I wish I could transport you over the leagues of space dividing us, and show you our "way of life," our new home and surroundings Our housekeeping is creditable, I assure you, for neatness and cleanliness. The presence among us of our now sedate and home-keeping cat gives it a home-like and domestic appearance. I often wonder whether that demure-looking canine[52] specimen has any reminiscenses of its past life, to be called up in the long evenings, when, stretched behind the stove in all the luxury of expanded legs, he muses on the past. He is an entirely different animal, now, from the gaunt, lean and erratic character he was when I "plucked him as a brand from the burning," and conquered his impulsive disposition to throw himself from the wagon into the ambient air, careless whither he went, in his disgust of life. He has become changed by living in a "respectable family," and acquiesces in all our rules, keeping good hours, and showing no disposition to return to his previous vagrant and equivocal habit. The only time I ever saw him manifest an inclination of the kind, was one very dark night when, supper over, and our attention called off, he stole out of the door and scampered off. He hadn't been gone many minutes before we saw him coming, like a streak of lightning, through the clearing in front, and tearing, with tail stretched out horizontally, stiff as a poker, into the open door, dashed into the further corner; when he immediately put himself in an "attitude of defence," raising his stiffened tail perpendicularly, with back curved like the letter *U* inverted, and spitting with a vehemence he had never shown since his exploits on the Pack Trail. What he had encountered in his nocturnal expedition, we never knew; we looked out, but could see nothing; and since that time he has been "all that we could wish." One evening, while seated at supper, again we heard a peculiar, shrill scream from the woods, and instantaneously, and as if shot, our sedate pet threw himself into the same state of rigidity he had displayed on that (to him) memorable night.

Conspicuously where my bed is placed, near the window opening out to the bay, is hung William's[53] minature, which he gave me at Cincinnati, on my way out. I regret much now that I did not bring *all* yours with me, but I supposed then I could at a thought recall all your several features at will. It grieves much to confess that I cannot at all times do so, and this lapse of memory is one of the most singular and strange consequences of the "Oregon trail," the body and mind, from the hardships and exposures attendant, becomes racked to an intensity that renders the memory capricious, long after the physical man has recovered from the tenison, the mind and memory are still subject to its own caprices, which at times become very tantalising. When I want to remember, I *cannot*, by the utmost effort of concentrated will, and at other times they will "come like shadows, so depart," and instantly will be recalled all I had vainly endeavored to recollect before. These freaks of memory are fast disappearing (thank Heaven) in the regular daily duties and calm and quiet life I am now living, but they were very trying while they lasted. Sometimes I forgot portions of the route I have come over, which will return again to me when least expected, "on household cares intent," or in the deep silence of the dark woods, only broken by the sound of my axe. The other day, without wish or will on my part, flashed upon my memory, illuminated by the same bright sun as shone over the scene when I last beheld it, the wild passes of the Blue Mountains. I saw distinctly, *and as if I saw distinct from the picture, and had no part in it, my own self*, riding down the ridge that led into "Lee's encampment."[54] The tall pines stood out grandly upon the crest of the mountains, in clear relief against the skies beyond, and down the deep ravine, over which the road seemed almost suspended, the dwarf oak and gnarled cedar twisting their jagged roots into the rocky walls, a solitary "foot packer" talking to me, and visibly again came before me the wild, weary, haggard and God-forsaken look of the poor creature, as he begged me "for God Almighty's sake to give him some thing to eat, as he had eaten nothing for two days, and feverish and sick he was afraid he could go no farther;" while I (God help *me*) had just broken my two days fast in "Grande Route [Ronde] Valley," and thought with a remorse (I never thought a few mouthfuls of bread would occasion,) of the huge meal I had eaten, and this poor fellow almost starving! I could give him nothing but a few choking cheering words of comfort, and a little money, with the hope that he might soon overtake some trains who could better spare provisions than those he had passed. In my dream, I remembered well how I looked back as I journey on—*even as I had then*—when far down the descent, to see him, high up, slowly plodding downward. These dashes of recollection are so vivid, coming upon me at such unexpected time, that I wake from them in mid-day, with the bright sun shining on me, with a start, and fancy it "was not *all* a dream" and well do I know they are *not*, but were incarnate realities, *happily now all over.*

Tuesday morning, and we are laughing heartily at Mumpford. He had dropped his axe down the bank and went down to get it, but directly came tearing up the bank as hard as he could—"a whale, O, whale"—we all run out to the bank, and looking out in the bay, we saw something floundering in the water, "backed like a whale," looking indeed "very like a whale." They were a couple of porpoises engaged in a morning's gambol. The "whale" will be a standing joke against him "for all time." {We of Pennsylvania and Ohio were, for purposes of porpoise recognition, as far inland as Iowa, but with a fine reticence, we had not shown our ignorance, and Iowa had. Probably the difference in longitude was shown in the incident; we were nearer the nonchalance of a higher civilization.}[55]

We have a couple of large bald-head eagles continually hovering near us. Their favorite rest is on a huge fir on the margin of the Bay. I tried my rifle on one the other day, but though I cut off a twig close beside him, he flew off unharmed with a shrill scream. They are tremendously large birds and handsome, but have bad habits—carry off chickens *when they have a chance*, and as I some day expect a legal and classically named friend will send me a "Shangai,"[rooster] and intend crossing the breed with a Flamingo, I want no poachers in the neighborhood.

The crows "who hereabouts do dwell," I have "tabooed," and taken them under my special protection and patronage. The very novelty of their being so tame as to allow you to come, while feeding, within a few feet of them, is so complete a reliance on your forbearance and an appeal to your confidence, that no man with a single "bowel of compassion" would violate it. Besides, they are a benefit—they are the scavengers of Oregon, removing all the offal, etc. There are thirty or forty of these "gentlemen in black," who occupy the peninsula in common with us, and if it was not for their incessant, monotonous cawing, they would be beyond reproach. I admire the rook or raven's cry—it is sonorous and full of melody. The crows are the "Corus [Corvus] Picas," or Fish Crow.[56] They have (since their assumption of scavengers for us,) forsaken their old business of clamming to less fortunate competitors. All day long I used to watch them, when the tide was out, picking up the muscles or clams which the receding waters had deserted, and, flying high up, drop them on a stone to break the shell—and, so accurate was their measurement and eye, they rarely missed their aim. If the first time did not succeed, they would ascend higher, and, dropping it again, try till they broke it[.]

{The Crab following up the tide upon the shelving shore, to anticipate the all devouring water, may use such marine cuss words as he can in a briny way command, when he finds Corvis Piscus above his knee joints in the ripples, appropriating what he assumed to be his exclusive provender, but he makes no close protest; how he knows, we may never know, but it is within his limited knowledge that Corvis is very fond of shellfish, and he conducts himself accordingly.

{Their vocabulary seems very limited, I have not fine enough ear to define whether by intonation, prolongation, or accent, they make it more elastic, but to the human ear it is as monotonous as it is persistent. I am satisfied they have differentials that we cannot detect, because what seems but a repetition of the same sound to us, evidently conveys more than one meaning to them.

{They are very quickly adaptive. I have various times heard a great clamor, and seen the large majority expostulating with what I supposed from their action, and his apparent lack of ease, to be a stranger, and at last seeming to find the greatest point of their argument lie in their bills, drive him before them vigorously. Only two or three at first made this last conclusive argument, but I noticed when these got the interloper on the run, or rather on the wing, every one took a peck or two at the unfortunate fellow. In my interested but very limited observation, I have observed there are some salient traits that are common to all animals.}[57]

They find it less trouble now to prey about the house—it *pays* better, and requires no taxation of ingenuity. {From a Crows point of view, there is often a generous residue from our Clam Chowders, with a flavor unknown to the raw bivalve. Its temporary companionship with edibles unknown to Crow gourmands, and the absorbence induced by fire, gives them an article of diet as rare as it is delicious, and I hear them expressing great appreciation thereof.}[58]

In despite of facilities being so good for farming here, so few are the settlers that provisions are very high. Potatoes, as I write you, are worth \$2 ½ per bushel; Wheat \$4; Flour \$40 per 100 pounds; the barrel I bought for \$40 came from England, and was brought here from Victoria, Van Couvers Island.[59] It is, as you may suppose, old and somewhat musty, and at home would not sell for \$2, but it answers our purpose, and I contrive to make some tolerable "soggy" bread out of it. When Spring comes I calculate to depend on nobody for vegetables, at any rate.

The enterprising company entitled "Adams' express," are about to establish a branch of their establishment at Olympia. When it is effected I will send you a barrel of salmon, and some of the Indian curiosities procured from Queen Charlotte's Island." Among them is a full length figure, on black slate, of the ships cook[60] they took prisoner, who being an Ethiopian, and possessing a good, honest face, would do admirable for a representation of "Uncle Tom," in the next "cabin" exhibited at the Horticultural Fair.

Saturday—our old Indian has been here again, this morning, bringing us "Hi you"[61] fish and a couple of ducks, and as he entertains a high opinion of my cooking, he has invited himself and a friend to dinner, and we expect a jolly time. I am up to duck-stuffing now, and am particularly expert and choice in gravies. To give the preparation a feminine look, I have ventured (in Ensign's absence) to don the sun-bonnet, and so, whenever I am at fault in some delicate preparation, not being able to recollect the "last mysterious pinch" of something that gives a flavor to the whole, I pause and try to imagine I am a woman—and recollections of what is wanted comes instantaneously and natural. I really think my talents are "hid under a bushel." If I ever return to the States I will request the first strong-minded woman I meet to go out and take my place and I'll cook for a subsistence—I would make the fortune of any restaurant.

{Oysters having a delicacy all their own, will not accept companionship of any but the best butter, but roasted in the shell, with no salt save of their own furnishing, are satisfying to the palate, as an after dish, but the wielder of an axe cannot hold enough to keep himself thoroughly content between meals. The Clam, being more sturdy, may accept an addition of a bit of pork in case of a stew, but he loses somewhat his personality, and has sufficient succulent resources of his own if confidence is given him, and he is not taken out of the shell until he comes to the table. My men were in no way epicurian[62] and wanted quantity, rather than quality, and got what they wanted.}[63]

I write this letter in the intervals of my duties; dividing my time equally between writing a little, watching the pot boil, and an occasional out-look on the fir, to see if Mr. Eagle is there. A huge raven has perched on a branch of a cedar near, evidently regaling himself with the fragrance that permiates the atmosphere around, from my boilings. He is only twenty feet from me, and could easily be brought down,—he is safe for me, though—I like his note.

We have found a prize! A ship's boat, near the point of our claim, drifted doubtless from some vessel, and has been there apparently two or three months. We intend (if it is not claimed) examining and repairing it, and going on a "voyage of discovery" to see the extent of our domain, and afterwards extend our survey all round the Lower Sound—will keep notes of our voyage, and if we discover any unknown Islands or tribes will send you due account.

Canoeing to Olympia

I must draw my letter to a close; must go up to Olympia this afternoon, as we are out of beef, and it may be I may receive some letters too.[64] If I do, I shall acknowledge their receipt in this—will have a tough time of it too; for the gulls have been screaming all morning—sure sign of a coming storm. I shall be until after dark making the trip, too—rowing both ways against the tide, and shall have no doubt a glorious ducking. But I have accustomed myself to these annoyances, and the time has long ceased when a rain occasons any bad consequences, save the inconvenience of the thing, and an unpleasant cense of being a "moist body." {I go up in my little canoe. I have mounted a little squaresail that does finely with a wind directly astern, but it is seldom so complacent, and on my return I almost always, if the tide is against me, try to quarter the wind a little to make my course, and as I have no keel I make great leeway, and as I *will* keep as near my course as I can hold the canoe, I almost always capsize. I cannot sail if I have any freght that would sink or damage by water, and I can only indulge in a capsize on a homeward trip, when I can change my clothes. The canoe has a long overhang at the stern. Now it is not possible with this little canoe to get aboard over the side, but by treading water at the overhang, and raising the stern high, and surging it back several times, the greater part of the water can be thrown out, and by crawling in over the extreme end of the overhang, the canoe can be entered. As one has to sit down in the bottom of the canoe, it is not practicable to get all the water out nor is it necessary, for you are thorougly wet from going overboard.

{As the canoe is half water logged, the exercise of paddling keeps you well warmed up, and in the salt water you do not get cold. You *can* in such stormy weather if you are content to use the paddle only, come and go without being drenched. You are sure to get damp, for with only a few inches of freeboard and a high sea running, you are bound to take in enough water to put an inch or two in the bottom, which you sit in.

{My own experience with water leads me to prefer a complete to a partial wetting, if I have to remain wet very long. Of course the little canoe is not adapted to a high sea, and the indian does not use it in such weather, not as a question of safety, but comfort; it is safe enough, thoroughly water logged, it will support a man. But there is a heap of joy in buffeting with the waves, and great comfort in a dry suit when you get in.}[65]

Sunday morning, at Olympia. I *did* come up last evening and had a terrible time. Old Boreas was on a frolic that night, certain. The storm came on when near the beach, and increased in severity as I landed safe.

"It grew a cold, and hail came down,
And a sharp numbing breeze,
As if from desert continent
Of ice in Arctic seas."[66]

I did not return last night, but remained with a friend till this morning, liberally rewarded for my journey by the many letters received, not the least welcome ones being from my little nephews, in return for which, when I have a chance, I'll send them a companion, a young So-wash,[67] (tell them) with a head so flattened they can play marbles on it. This letter is already too long for me to answer in it their contents, besides I have not yet realized them—never *could*, in my excitement at *first* reading—containing as they do pictures of far-away and home scenes and—

"—Truths sublime,
That need a quieter home and calmer time"[68]

to enjoy them, than I now have. I'll float gently down now *with the tide*, and read them as I go, and as soon as I can, will write you a little postscript (say of two or three sheets in length,) meanwhile—

"With such a prayer on this sweet day
As you may hear and I may say,
I greet you from the "Far-away."[69]

Letter to the *Dispatch*[70]

ALLEN'S CLAIM, three miles below}
Olympia, O.T. March 27, 1853

Messrs. Editors:—I wrote you last, a scribbly little letter from Portland—in which I threatened (doubtless to your great dismay,) to write you again upon the "mellowing of occasion."—I have now been upon the shores of Puget Sound some four months, and until of late days have felt but little of the "caoethes scribendi,"[71] beyond the writing of "home letters," which never seem a task to me under any circumstances—and which I have written I believe in all postures—lying under the wagon, cramped up *in* the wagon, on the horn of my saddle, &c., &c., and only need, to complete the series, to scribble a few lines standing on my head!

I desired, earnestly, to have written a letter in time for the coming emigration, in time to have been published for their benefit, and had drawn off the heads of all matters I wished to speak of: outfit, wagons, cattle, place of starting, &c., in the hope that a few words from an "eye-witness," would save those of my friends who designed taking up the line of march for Oregon much of the annoyance and "disagrement" that my verdancy occasioned me.

I should weary you with detail of hardships, if I were to recount the incidents of our trip from Portland here; —coming, as we did, (in a season when all travel is suspended,) upon a "pack-trail," hard enough at any season of the year. Suffice it to say that I began to feel somewhat of the dogged endurance experienced from Fort Belle to the Dalles of the Columbia. I felt so worn down, that I could not summon up courage to do anything, save eating most monstrous dinners and bathing in the luxury of idleness. All this has grown "small by degrees and beautifully less," excepting the dinner part—which has, on the contrary, increased to an extent quite alarming in a country where flour is $40 per bbl [barrel].

I have (as you see by the heading of my letter,) taken a claim some three miles below Olympia, at which place my friends are respectfully requested to "buck up" in the letter line. It is upon the west side of Budd's Inlet only half a mile across, and bounded again by Ell's Inlet, which contains the finest oyster-beds upon the Sound. I "put up" at Quincy Brook's when I first came in, and there remained until I grew strong and energetic again; when, in company with a young Ohioan, (Shaley [Shirley] Ensign,) I crossed the bay one Saturday night, lit up a rousing fire, wrapped up ourselves on our blankets, and had a glorious sleep, awakening in the morning "landed proprietors," such being the law of practical Free Soil.

We have put up a snug little cabin; through the windows of which, as I write, the warm sun comes streaming, and the south wind brings in the murmur of the waves upon the beach, and the screaming of the sea-gulls as they swoop down upon the foamy water. It is doubtless "black March" with you—with us "the time of the singing of birds has come," and the woods filled with the "light of the new born verdure" gives us full assurance that our Spring is here.

Ah! this Oregon is a beautiful country! I know of no place where a man could live with more pleasure or with less care for the morrow than here. The government recognises his inborn right to the soil—"man's heritage," and it gives him, if single, 160 acres of land—only binding him to remain upon it for four years, cultivating it some little every year. If he wish a location where he can look out upon the "world of waters," here upon the sound he has before him where

This rough map shows the location of Allen's donation land claim north of Olympia. The map is by an unknown surveyor, but was probably a draft for a more refined land claim map prepared for the General Land Office survey. Northwest Microfilm Collection, NW maps, Thurston County, WA, 912.7977 NW, T 19 N R 2 W. Washington State Library.

to choose—either a place where the falling leaves have enriched the soil, mayhap for centuries, or where the prairie stretches back almost from the very beach, inviting culture—or where the narrow belt of timber lines the margin of the bay, and merges into the beautiful prairie back.

An "outlook" upon the bay itself, with its flowing and ebbing tides and ever-musical murmuring of waves, is of itself—to one born inland—worth many disadvantages of location. Here, as I sit, I look across the bay, some two or three miles, (narrowest at this point,) and upon the other shore see the little openings which contain the "wee" cabins of Brooks and my kind neighbor, Mr. Wylie,[72] of whom it is only necessary to say he subscribes to the *National Era.*[73] Upon the shores are Indian camps; while, far beyond, Mt. Rainer looms up clear and distinct in the clear sky, ever snow-crowned, not unlike the huge giant guarding the country, which the Indians suppose it to be. It is yellow and golden in the bright light, seeming like a great mass of molten metal. Some day its solitude will be broken by the sound of the miners' pick and shovel, and many a hearth at home will be gladdened by the success of the unkempt personage, who is only a "miner" here, but in a far off land is a son and brother—or saddened by the news that the giant peak, with its pure mantel of white, overshadows his last resting place! There will doubtless be gold found all through the Cascade Range, and already the Portland papers announce the discovery of gold upon the headwaters of a stream heading upon the east side of Mt. Rainer. It is almost due east from here, and beyond it lies—home!

All this you will say is very beautiful, and would look well in a picture—but a man don't eat with his eyes. Right—but that is the *poetry* of the picture; and, as even a poet lives on cabbage, here is the practical portion: Should a man take a prairie claim, he can go at once to work putting in crops, only he must possess enough money to keep things moving in the meanwhile, or in the intervals must "hire out," helping some wealthier neighbor to clear his claim or working at wood chopping, at which even such a town bred diminutive "atomy" as myself got $3 per day. Should he take a timbered claim he can do as we are doing—get some four or five together, and get out hewn timber, which has always sold heretofore for $10 per 100 feet, costing $3 for delivery by cattle—putting in crops as he can spare time for clearing.

Potatoes are the most profitable, because the least troublesome crop. They now are, and have been ever since my arrival, $2.50 per bushel. The ordinary crop is 300 bushels, realizing $750 per acre. They will doubtless be lower next season, but they will always command a good price, as there are none now shipped, and California will always cause prices to rule high. The farmers here never pay any attention to them after planting, and frequently the second year a crop of "volunteers" spring up, almost as full as the first year. Some of the farmers are planting now, it being their customary time, while others do not plant until the last of June. Onions are $4 a bushel. Turnips here don't sell by the bushel, as you could not get more than one in a measure.

There are splendid openings here for dairies. Cheese sells at Portland at 50 to 75 cents per pound. Here, *it don't sell at all*, for the simplest reason in the world. Butter is scarce here, at $1 per pound, and yet, on the broad prairies, any number of cattle could be kept, year after year, with no outlay for feed, save the cutting of one stack of hay, for fear of such a severe winter as the last—the severest known her for seven years. So says old "Pere Ricarde,"[74] at the "Mission," upon referring to his journal. California will always furnish a market for butter and cheese—and in our growing trade with China, there is no telling what a hole these long-queued Celestials may eat into our products. To a man of capital, the milling business offers inducements seldom to be met with. A steam sawmill upon the Sound is of itself a fortune. The hewn timber we are getting out is taken to California, and sawed into lumber there, because vessels cannot get cargoes here—mills are so few in number. At the falls of De Shute river, above Olympia, is the only saw-mill upon this end of the Sound. It runs by water power, and is now rented at $840 per month. The whole concern, built when labor was much higher, cost $5,000—thus paying for itself in six months.—Lumber has sold at $50 per M., and never less than $23, with a steady demand much greater than the supply. The lumber is all sawn from white, red and yellow fir, and cedar—which is all the variety of merchantable timber we possess.

With all these advantages, poetical and practical; clear, health-giving air; beautiful scenery and such reward for labor, you may ask, what under heaven does your country lack? To me, simply this: last night, inditing a letter home for one of our boys who cannot write, he bid me tell them all about the "big trees," "salmon," "deer," "porpoises," beautiful climate, high price of labor, &c. &c., but, winding up all, he bid me say that *it was not home*, after all. He "has" me where Portia had Shylock. So, some of these days I shall "gird up my loins and depart hence," and in the bustle and littleness of city life, remember when the "lines were cast in pleasant places," and I might have lived as man should live, and "would not"—because, "ever, through all, the burden of my most anxious thought kept drifting, like a sea-bound river, Homeward!"

Yesterday was mail day—and I loaned my canoe to one of my neighbors to go up to Olympia, inquiring also for me. I was clearing around the house, chopping most lustily, but keeping a bright eye up the bay, at the imminent risk of my shrinking legs. The little "dug-out" appeared, a speck no bigger than my hand, but came steadily on; I, like a hare, resisting the inclination to rush up the beach to meet it. It came in, at last, and Eureka, what a mail bag! Thank the Lord, if I am a "remote circumstance," the folks at home have not forgotten me. Letters, an infinitude of papers, and last but not least, a "hull heap" of *Dispatches*—for which, as they are done up in editorial style, I presume I have to thank you. My little shanty has quite a *litter*-ary appearance. I also received letters from Resser, who is about to start a tinnery at Marysville in connection with his late employer. Carnahan is at Jacksonville, in a store. McClure making shingles in Umpqua Valley.[75] Dr. Eggers is presiding over the culinary department of a number of wood choppers, on the claim below me. Some of these times look out for some "Ballads of Chinook," in the original Indian!

Last Sunday, while busily engaged in melting out some pitch from the bark that adhered to it, ("dissolving the union" in quite as harmless a manner as "the gentleman from South Carolina,") an enormous bald eagle was foolhardy enough to alight on the dead cedar overhanging the bay. "Emblem of our country," say you—"oughtn't to shoot it." But, look you, the "emblem" *will* "annex" chickens—shouldn't mind the annexation, occasionally, of a stray weakling of my neighbors, ("a-la" Mexico, that, all right!) and, if it was only to support the old "institution," shouldn't care—with the expectation that time would die it out—but, for aught I know, they all go to rear up a fresh brood of young eaglets. Only a day or two ago I saw that same old villain in an "United States Marshal" kind of a way,) snatch a duck from amid a "large circle of acquaintances," and carry him off—the Lord knows where! The skin of that "peculiar institution," as it stands tacked up against the side of my shanty, measures eight feet from wing to wing. The old rifle is worthy of the guarantee given it by Capt. Harry Brockett.

Night is drawing down (I was going to say,) its "sable curtain," but it is moonlight—"light as a feather." I, as "head-cook and bottle-washer," must see about supper. *Salmon* and potatoes for dinner—variety: *potatoes* and salmon for supper—after which a pleasant re-reading of home letters and papers—and then

> "A sleep—
> Full of sweet dreams, and health, and quiet
> breathing."[76]

Yours—under the blanket,
EDWARD J. ALLEN

A Logging Contract

ALLEN'S CLAIM, near Olympia, Puget Sound, Washington Ter., May 22, 1853.

When last I wrote you, I gave a somewhat brief account of my "settling down," building our house, the result of a week's effort in clearing our claim, and amount of stock of timber we had cut for sale, squaring the logs so cut at so much a foot, which are then taken to California, (from the want of mills here,) to be there sawed into the required timber.

We continued at this some time, gradually clearing our land and accumulating capital *in squared logs*. About this time the demand fell, or rather there were no vessels in the sound to take it away. By the time we had cut to the amount (at the least price ever yet sold at) of $1,200, the men became impatient; many others engaged in the same business had been compelled to cease, from want of money to go on; I had managed to keep us all in provisions pretty well, which were all paid for in cash, and being anxious to retain them, as they were all "good men and true," went up to Olympia, and adjacent settlements on the sound, to get a cash contract for getting out timber, pending the sale of what we had cut to sell in San Fransisco, to which place I had written to charter a vessel, and was fortunate enough to find a person who had a contract to get out 5,000 feet of unusually long timber; the very kind prevalent on "Allen's Claim"—*his* contract was 15 cents, I took it at 14—$200 to be paid in a month, balance when done. I got a barrel of flour, at $20 (he was a merchant,) to seal the bargain, put it in my canoe, and "klat-awaid"[77] down with it to our peninsula.

There were only two of us in this job, as one of the boys had taken it into his head to make a little tour around the sound. I told our men that it was a low contract, but I had taken it to keep all together; should not reduce their wages, but expected we would average 200 feet a day. Our first day's operations were 160 feet, only, on which, however, the men made $3 and Ensign and I $4 each. The succeeding days were better.

Perhaps it may be thought that this picture of Oregon Life is not a very inviting one; the ever-recurring day labor—but it *pays*, in more senses than one. Every tree cut down will pay for seed, and in the place where it once stood, may wave the golden grain, ripening, "reeling and blushing" to the sun, on your own acres, mark you!

Nor is it so *very* laborious. I go out daily with them; and, returning some time before meals, *cook for seven men*, including myself—and though sometimes in the evening I come home somewhat fatigued, it is not the "kind of tired" (as the little boy said,) I used to experience when returning from daily avocations a year back in the "deavin dorismoe[78] town." *Here* the head is clear, though the limbs may be wearied—for the surroundings are vastly different. Months ago, the Oregon summer sprung upon us with a bound and our daily walks to our "field of operation" (getting to be a tolerably good sized field now!) through the silent woods, are made pleasant by the greenness of the grass we tread upon, the fragrance of the many-hued flowers, and multitude of berries peculiar to this God-favored country, while

"Sea-like murmurs, deep and low,
From the old trees rise and fall;
One seems to feel the ebb and flow
Of the solemn heart of all."[79]

Politics

It may have been a source of wonder, to many who have read these sketches, that so little allusion has been made in them to the state of political parties in Oregon. There is very little *can* be said—at least as regards the Northern part, Washington Teritory, where I have the happiness to live. There are any quantity of politicians, I am sorry to say, even here; disinterested patriots, willing to "serve their country" in anything, except, may be, viewing and cutting "Cascade roads!"—but the respective force of the opposing parties is, as yet, unknown, and this uncertainty is a matter of painful perplexity to the "fence-riders," who go in for "men, *not* principles."

I am no "politicianer," myself, and never contemplate becoming one—which, by the bye, reminds me of a conversation a firm and fervent friend of mine, and a veteran politician beside, indulged me with, on the eve of my leaving for this "uttermost part of the earth."

"You are a young man"...."Yes, Sir," (deprecatingly, but adding eagerly,) "I'll have a vote when I arrive,"......"I mean, comparatively young".... "Yes, sir!".... "It's a new country you are going to." "Same age as other countries, I believe—but less known, maybe." "My meaning, Sir." "You'll make your mark, doubtless there." "Expect I will—chopping trees." "No matter at what—but you will or can—but one thing I must warn you against; your opinions are not formed yet—political, I mean." "Oh yes, they are." "Well, they *should not* be—or at least till you know what is going to be the *prevailing party* in your new country."

I am afraid my venerable friend's advice was lost upon me, for my instinct and convictions are apt to leave me in a minority on all great questions of political and social reform, and my sympathies and voice are with that party who call themselves "Free Democratic;" maybe however that in this new country, rapidly filling up by the free tide of emigration from Iowa, Illinois, and (I may say) Missouri, rolling over the prairies, I may not be so long in a minority as my fellow thinkers in the states. My impression is that, when the parties are brought out to front each other, the Free Soil influence will not be so despicable. A club of 32 subscribers to the *National Era*, from Olympia and thereabouts, is no slight evidence, in my mind, of the probability of this hope. It is one of the proudest thoughts of my life, that my first political act should have been the signing (with ten others,) of the memorial expressive of the wishes of the people of Northern Oregon, that caused the creation of one more territory—some day as a State, to send her Senators to advocate Free Speech, and *American* measures. Washington and Oregon Territories are *each* large enough to make *two* States, larger than many of our largest. Nature, indeed, (by the Columbia River and Cascade range of mountains, running transversely across the country,) seems to have created natural boundaries for four.

I am writing this letter at intervals, intending to mail it when our mail comes in—every fifteen days, when it comes at all! which it sometimes *don't* for five or six weeks; latterly it has been regular; as I have nine days ere it arrives again, and I write in it by spells, it may chance, ere it is finished, to reach a goodly size.

Our cabin presents rather a bachelorish appearance this rainy morning. One man is setting in his bunk, sewing up a rent in his breeches; a second by the window, indulging in the unusual luxury of combing his head; a third cleaning out the bone of the wing of an eagle, to make a pipe of—while I am seated at the opposite window, an out-look on the bay, at our little pine table, writing this letter. I gave you, before, some general ideas of the interior appearance of our cabin; the "bunks" are built upon the side of the house, after the manner of steam boat berths; windows entirely innocent of glass. Our beds are composed of fir branches, with a little straw over them, and an Indian mat over all—and our sleeps are refreshing.

A barrel of salmon in one corner, nestling to a barrel of flour—between the windows a few rough shelves, containing our library, a very f[ew] books, some files of *Dispatches* and *Tribunes*, and some cherished copies of *Knickerbocker* and *Harper*; a corner table holds an array of plates, dishes, &c., and a respectable pile of newly baked cakes, ready for supper, pending eating which I write this letter; a puff of wind off the bay waves before the door three shirts, suspended on an adjacent tree—immaculately white, only "done up" this evening!

I am frequently interupted by the boys discussing Spiritual Knockings—in which Ensign is a firm believer, and relates some singular things he has seen and heard at Olympia; which, however strange and marvellous, granting them true, are singularly subversive of all opinions as to their spiritual origin, so utterly at variance are their revelation of a future state with my conception of the "blessed land." All they have as yet "rapped out," in *that* neighborhood, is not calculated to elicit much devotion—nor the questions asked and experiments made calculated either to raise in ones estimation the parties who are actors in such "Olympic games." The conversation has changed, while I write, and Ensign is now absorbed in the relation of the doings in his Ohio home, "Bees," "Apple Parin's," "Quiltin's," "Huskin's," &c.—all of which he saw, part of which he was"—incidental reminiscences, and mentions of the heroes of that locality; "Chump Williams," "Dad Smith," &c., are called up, with a pathos that shows how lasting are our youthful impressions This may doubtless seem all very commonplace, but were you one of the spectators and listeners, as I was, and your heart also busy with its own memories, as mine was—you would have listened with as much interest to these apparently trifling details.

"That heart, methinks, were of stony mould,
Which kept no cherished purity of earlier, happier
hours,
When life was fresh,
And love and innocence made holiday."[80]

Making a Canoe and Other Native Information

We have become somewhat more familar with our Indian neighbors than when last I wrote you. A curious people! the last link in the Indian descent; tho' not as degraded as the "diggers" of California, they are equally helpless or "shiftless," affording the most ample field for *genuine* missionary labor of any corner of the green earth.

{A curious people, and from our standpoint, difficult to understand. We have had centuries of a cumulative tendency, our wants have multiplied, and to supply them we have increased our strenous labor.

{The indian has no more wants than he had ages ago, and feels no more necessity for labor than he did when the world was young.

{To us he appears to have gone back, where he has only been stationary. In a way he has had a happier life, in that he has had less care, but we have no desire to take up the ease of his life, in exchange for the harass of ours. [A]nd because we would not, we are unable to judge him fairly. If we could but compromise, and give and take, we might both be happier, for a time, but only until we discovered that it is but a low form of happiness that is based on stagnation.}[81]

There has been, until very lately, a large camp of them a few hundred yards above us, on the beach. They were living previously on the other side of the bay, but wanting a canoe, as we have the handsomest timber on the bay, they came over to construct themselves one. An Indian always consults convenience in all he undertakes involving any labor; he would not paddle to and from his work, as that would be too systematic, too much like work—so he tumbles his lodge squaws, and papooses, in his canoe, and over he comes, sets up a few sticks, places his mats (curiously made of rushes,) against them, lays some over head for a roof, "enters in," having erected his house and taken possession in half an hour!

One day when out cutting down trees, we thought we heard the echo of others chopping beyond us; Ensign went up to see, and found the Indian cutting down a cedar. He ran and hid, leaving his blanket behind—but, upon Ensign smiling him back, he reappeared with reluctant and lingering steps. Ensign could not talk "Chinook," but by this time I had come up, and told him to "fear not, cut away." I thought it a hard case if they could not be allowed to cut enough timber to make a canoe, on their own land.

{You see, it seemed a little rapid, because we had set up our lodge but a few weeks ago upon land that up to that date had been open for centuries to all indians, that we should with a few weeks occupancy, utterly bar them out.

{Their ownership, unlike ours is not individual, but is general and is vested in the whole tribe, not as a matter of record, but with centuries of occupation. If we had by treaty, bought out their rights I should have felt that I had the right to consider the indian a trespasser, if I so elected, but I most certainly would not have so elected.

{He was heartily welcome to his canoe cedar, and I was glad to tell him so, and with no assumption of generosity.

{You see even if the Government had bought him out, and paid what was agreed upon, it would have been at the rate of about one hundred acres for a nickel, and the Government would at once show their idea of its value, by placing it in the market at the minimum rate of $1.25 per acre, an extremely low rate, which I got the benefit of.

{Quite likely there would be given a reservation of land to be his forever, which did not include in it what the indian required, probably it would not include trees that were fit for the making of canoes, and with the indians ignorance of our different ideas of ownership he would not, when he signed away his lands, understand this.

{As a fact the word "given" which I have used is an error, it was not *given*, it was, as the name "reservation" signifies, a portion not included in the sale, the century's old title, remaining in the Indian.

{As a rule, in these treaties, the necessities of the red men were not very well guarded, not so much because of fraud, but from ignorance of their wants, and a large indifference.

{In the case of my canoe man, from his standpoint I was the trespasser, as there had as yet been no attempt to purchase his lands.

{When they are purchased, it is really at best, a jug handle affair, it is in a decided measure, compulsory.

{I am not criticising this last, it has to be so, but it is well to keep in mind, the conditions, and exercise some consideration.}[82]

What an Indian *does* know, he knows well. He had selected as good a tree for his purpose as could be found; he was a week or ten days finishing it, working at it in a desultory Indian kind of a way, just as he felt in the humor. As I repassed to my work, daily, I noticed his progress. He first stripped off the bark and then cut away the sap part of the tree, getting to the hard and tough part of the wood. Then, with no guide except his eye, and with no tool except an old rasp, he had sharpened and fastened something like a guage, in a root-like handle, he fashioned a beautiful canoe, about 15 feet long and a little over a foot wide—of the most graceful shape imaginable. The greater part, inside, he burned out, finishing it with his guage. Then making a huge torch of pitch-fir, he scorched it all along until it was well charred and burned, all the time putting braces across, widening it gradually, till it was two feet wide, beaming it with cedar, fastened with withes; and, after scouring it out well with rushes, it was finished—a model of grace, weighing seventy lbs., and capable of carrying eight hundred.

{When they make a large family canoe, they give it a high prow, always painted with the same insignia, resembling a conventional eye along the top of the canoe, they will insert what look like rough pearls, at first glance, but are the hinges of the clam shell.

{The cedar is selected because it is so peculiarly straight grained, is very enduring, works easily, and has very little sap and does not warp or shrink.}[83]

Our little canoe had been broken, by allowing the waves to dash it on the shore, and I got him to repair it. It was cracked along the bottom; he hauled it out, let it dry in the sun two days, turned it up, cut a strip of cedar bark, as long as the crack and about an inch wide, pounded up a lot of cedar bark, and (rubbing it with a mixture of some stuff they take from a kind of fish they catch, something between a flounder and alligator,) caulked it tightly. Then laid the strip over it, and—boring holes in each edge, through the bottom of the canoe—pegged it over; and, covering all over with a thick layer of the mixture, let it dry; it was "as good as new."

A day or two since, one would imagine there was an unusual excitement among the squaws; over a dozen or more were engaged in digging but they had been engaged thus busily every day for two weeks past gathering and drying clams, to be ready to make their annual migration to the "hill country"—now that the berry season has set in. The camp of our Indian canoe maker, above, was perfectly strewed and strung around with clams, smoking and drying, ready for use. The largest size grow at the end of the bay, and they come here to dry them.

{A plot of an acre at most with but little attention would produce all the berries they would need, but they pack up all their stuff and go maybe, a hundred miles to a berry ground. An acre in potatoes, of which they are very fond would yield enough for two or three groups, and add something of change. Here and there, Indians in a haphazard way raise a few such potatoes as we leave in the hill, as not worth the labor of gathering, but very few of them do even this. In the Spring I notice they take the young shoots of the blackberry and stripping off the briary outside eat the tender inside, as we do celery, and seem to enjoy it very much.

{I suppose my indian needed another canoe to complete his transportation. He may have taken an additional squaw. If he had the goods necessary for purchase, there was no other economical reason to prevent him. You see their methods differ from ours, the "upkeep" of a squaw is negligable. She comes into the firm with a blanket and that lasts for years, its lease of life is not shortened by washing. The blanket, and possibly but not certainly some equally long lived leggings, constitute her whole wardrobe. The seasons bring no changes in her costume, there are no "Easter styles" and she makes no outlay at any store, the paint on her face is not procured at a drug store.

{The pigment is taken from the claybank with her own dusky fingers, and spread on with a breadth that is not restricted by any question of cost. She digs out of the beach, not only the clams for her own support, but such excess as may be needed, also the Camas[84] is of her digging and drying, and the berries, of her collecting, and the fish of her drying, her support is not a matter of cost to the man, she is absolutely a provider. So it comes that no after economic cost of the women is a consideration in the inclination to marriage, and the Indian marries young – and often, if he wishes to. As a corollary, if there is in any indian tongue a synonym for prostitute, I am constrained to believe it is of recent introduction, and the enrichment of the language is due to the white man.

{It is a curious thing to consider how much the acceptance of a fashion plate would, or might, have introduced these bad words into the indian tongue, and the appreciation of a "love of a bonnet" thrown obstructions in the way of marriage, and so have lessened the chastity of a tribe.

{I can hardly think that Eve's introduction of a simple fig leaf to what might be considered an utter absence of costume, would have worked woe, but you see her womans delight in beautiful things in form and color, came into play, and doubtless she commenced with cutting it "bias," putting "gores" in it, and getting at last to "trimmings," and then when some shrewd people began to manufacture for her desires, she began to be costly in the upkeep, and the man had to stop to consider, and marriage became not so much a matter of course.

{I wonder if that is the way of it. I am here at the beginning of things, where there is no wardrobe and everybody marries, and marries a good deal. And I migrated from a place where there was a great deal of wardrobe, and a lessening number of marriages.

{There may be no connection at all between these two conditions, and I am poor at analysis, and the relation of things, but some one who reads this, may think it interesting, and want to look into it. These "Abergoins" [Aborigines] start a fellow thinking in odd channels.}[85]

The Indians all have to buy their squaws not with costly annuals, albums, sleigh-rides or concert tickets—but with horses and blankets at prices ranging from $5 to $200, according to the demand, or value of the subject—without which the "old min is never friendly." Daughters are frequently married (if that sacred rite can be so called, as practised among them,) at twelve years of age—and valued at exhorbitant rates; a good squaw clam-digger will, in the purchase, absorb an Indian's small fortune!

{Riding down from Tumwater in company with Mike Simmons, the old settler before spoken of, I was startled by an emphatic exclamation and his quick dismounting, and dashing into the bush, his quick eye had seen a squaw hanging from a low bush, evidently so determined on self destruction that her knees were drawn up from the ground.

{Simmons cut her down, and I heard him talk some very strenuous Chinook, using some words that I had not known were in the vocabulary.

{He told me she was attempting suicide because of jealousy. Now it seems a singular thing to say, but I felt that life had more interest for the indian women that I had supposed, and I was glad to know it. It was something more than the mere drudgery it seemed.

{I felt like taking off my cap in the unexpected presence of that divine instinct which God has implanted in women, that enables her to secure the consolation of love, in seemingly impossible conditions.

{I was solicitous lest she would attempt it again after we passed on, but Simmons said she would not.}[86]

They are very anxious to marry them to "Boston men," as they call the Americans, and many of the old settlers, traders and Hudson Bay employees, have, in past times, taken them as, by this means, they secure also the services of all their kith and kin. The poor creatures are exemplary in their new relation, and seem anxious to have their relations used as much as possible to advance the interests of their husbands.

A neighbour of mine, a residenter in the country for over forty years, once a "born thrall," of the Hudson's Bay Company, since a farmer, and a man of substance and acknowledged reliability and worth, influenced by the iniquitous policy of that powerful company, in common with many others, dreaming themselves out the pale of civilisation, "took to themselves" squaws for wives. The time has rolled around when it has become a source of regret and mortification to them; but, to their credit be it said, that I have heard of no one instance where they have acted the unmanly part; living with them so long, doubtless has begotten an affection hard to stifle; children have been born to them they cannot repudiate; many of them have striven to anglicise their wives, making them presentable. The squaws, obviously aware of their position, and proud of their liege lords, zealously second their effort; some of them are very nice and clean about the house, very anxious to "do the agreeable" to any visitors chancing to come in, evidently proving that intercourse and contact with the whites, when exerted for their good, is as productive *of* that good—the converse of which produces so much evil, the *details and features* of which I *could but hint* in words.

{You see, in Indian ethics, the continuation of the alliance depends entirely upon the man, if for any reason he repudiates it, the woman goes back to her father, who as a rule makes no objection, as he can sell her again, if she is yet desirable.

{The man loses the dowry, or price he paid, the unwillingness to do so, is the womans only security. Pretty slim isn't it.

{I am a little ahead of my story in what I am about to add, but I do not want to leave a number of readers longer than necessary in a serious state of mind. When the Territorial legislature met, one of the early statutes made proof after a certain date of such living with any woman in that way, a common law marriage and legalised the offspring.[87] I think the statute was welcomed by almost, if not all, of the "Squaw Men," and I know of no case where there was desertion.

{These men made honorable amends and were respected therfor. The less said of many other men in these matters, the cleaner these pages will be.}[88]

The Indians seem to have no idea of the extent of their degradation. I have had repeated offers made me to "purchase a cook;" yesterday an old "Tice" came from above, with three or four, in a canoe—proffering to sell me any one chosen; one he recommended especially as a "wyash klosh oopet mamak muck-a-mucka,"[89]) an excellent cook," and cheap—only four blankets! which offer, it is needless to say, I most respectfully declined. We read and hear much talk about the "degradation of the Indian." "Wo be unto these, by whom offences come!"

Glancing out upon the bay, I see floating down what seems like a floating tree, with the green branches floating upward. This was the idea a group of ducks above us seemed to have—but a thin wreath of smoke curled out from amid the branches, and the report of a gun came booming over the waves. Two green-throated mallard ducks floated dead, "the others flew away;" a paddle emerged from amidst the mass of foliage, which gradually approached the dead birds, a head of an Indian protruded, and an arm, which "gathered them in."

I heard thunder this morning, very faintly; almost fancied it was the grouse drumming. It is but seldom heard here. The first time I have heard it since my arrival; had thought the Pacific coast would be much visited by thunder-storms but was mistaken.

States Mail

Last Sunday I walked up to Mud Bay, to hunt up a stream which was said to come in there; found it large enough for an overshot mill, but it would be somewhat difficult to construct a race. It comes in on the upper side of the claim we hold. Mud Bay, or Ell's Inlet, (the latter being the proper name,) is a most beautiful sheet of water, not so wide as Budd's Inlet. From that side of the claim, I have a lovely view of the Coast Range. The bay heads as high as Budd's Inlet. Somewhere up in that direction is Black Lake,[90] a large sheet of water, never yet explored—where the Indians from our neighborhood go every summer to fish; a good country around it. Dr. Eggers, who came over in 1850, had a claim there, but abandoned it, either because it was "too far from market" or impelled by that restlessness that has led him to abandon a half a dozen other good selections—one of which, back of Milwaukie, would have made him a wealthy man, had he retained it.

A neighbour of mine, a good fellow by the bye, complimented me one day on my letters. I asked him where he had seen any? and he related that being one day in the Olympia Post office, when a "States mail" was being opened, he saw a paper directed "To any clever fellow in Olympia," and thinking the editor had doubtless meant *him* but had forgotten his name, he applied for and received it. It was a copy of the Pittsburg *Dispatch*, containing one of them. The "devil" who envelopes the paper must have been "on a comical lay" that morning!

In a re-perusal of my letters, in print, I notice many comical and funny typographical errors. In one place they make me say "four *mules* brought me to Point Neuff Creek." I have grown somewhat fleshier, since I left weighing now 147 lbs, but *one* good "mule" can carry me yet. It should have been four *miles*. In another place they make me say that a "canon" we encountered in the descent of Lewis' River was "filled with reptiles," it should have been cataracts.

I saw a paragraph in a New York paper, the other day, giving an account of the fate of a party who descended after us, who were not so fortunate as we were—*only five out of twenty* arriving safely; they were probably the same party we left behind, preparing to descend after us. On our arrival at Fort Boise we were the first who had come down so far, as I could learn. There was a party that had started before us, from a lower point, which doubtless was the one we found wrecked. Others came down while we were ferrying, and went on farther—*none of whom were ever again heard of.*

It was a desperate undertaking—and I knew it *would* be, when I resolved to *attempt* it; but there was no choice left for me, and I have no doubt it saved me from being a cripple for life.[91]

Indian Fishing Practices

Quite a number of Indians have been spearing the fish that come to shore to spawn, and have brought us a number to "swap." Fresh salmon is also coming into season. Herrings [smelt], similar to Scotch in taste, but smaller, are abundant. {[Smelt] are in large shoals, the ducks by hundreds swimming over them, constantly diving for them.... The indian gets his smelt in a primitive but very efficient way. He takes a pole about ten feet long, and in three or four feet of the lower end, drives nails in so that they project an inch or two of both head and point, and swirls this back and forward in the shoal a few times, and pulls it in, and knock off on the gunnel of the canoe, the smelt that have fastened therin. In a short time he has all he can dispose of. He is no butcher as far as quantity is concerned, he kills only what he needs.}[92] Halibut are said to be plenty—though I have seen but few. The fisheries will be an important feature in the resources of this wonderful country, and already a large number of companies have been organised in California, and commenced operations, in drying, packing and exporting. In the Straits of Fuca the Indians capture whales.

As the Indians had been unusually successful in fishing, they had quite a large number to sell—and as an Indian considers it beneath his dignity to carry a bundle, like a slave, he had brought a squaw along to carry it. One of the principal grievances set forth in the Womens Rights Conventions—their exclusion from the many avenues of labor, would be useless here! for the "Lords of Creations," are remarkably liberal in that respect, throwing the onus of *all* the labor on their squaws, and opening wide every avenue to work. When an Indian is moving to the hills, in the "berry season," all his worldly effects are in a bundle, and bound on the back of his squaw (or squaws,) supported by a strap, which crosses their flat heads, above the eyes—so that, when leaning forward, it cannot possibly slip; and thus she tramps off! he following leisurely behind, with head uplifted, like a Shanghae rooster!

If I wanted to hire an Indian to bring oysters, carry shingles to the beach, or pile up brush when clearing—do you think *he* would come? No, *sir*—he would send his squaws!

Birds and Weather

.....(Thursday.) It is night—raining briskly—and the pattering of its drops makes music overhead as I write. The winds are blowing; and, though I cannot see out into the dark night, I hear the waves surging on the beach. Sea gulls scream above me, and I almost fancy I see them swooping out from the bay beyond. The tide by this time runs seaward, and the waterfowl are gathering into the bights and bays along the coast. Something, now, startles the ducks; and, through the pattering of the rain, a lull of the wind enables me to hear distinctly the fluttering of their wings, as hundreds, mayhap thousands of them, cross the Sound. The crows which (unmindful of St. Valentine's Day,) have neglected till now their duties, are fearful the wierd winds will destroy their half-formed nests, and scatter the carefully chosen twigs they have cawed in consultation over for the last few days—and call on each other, thro the darkness, seemingly in great perturbation; whilst I, remembering "the man in the fable" who understood the language of bird and beast, wish he had transmitted his knowledge, that "hearing I might understand."

Nothing seems to enjoy the storm and the night so heartily as the little wood-frog—which swells out its cheerful little note—and the owl, which "comes out jolly, under the circumstances," and is evidently impressed with the idea that the night has been created for its especial benefit—emitting such a sonorous "too-whoo" as makes the little wood-mouse shrink closer in its covert!

There was a bright-eyed coquettish little bird used to twiter about the house, and peer thro the window so knowingly and confidingly, singing such bits and scraps of song as to make us all conceive a liking for the little fellow—but I heard, one night, the sweeping of the great ogre-eyed owl, down toward the hazel at the spring; since which I have seen our little favorite no more! so some bright moonlight, when he ventures again to utter his dismal note near my residence, I shall protrude my rifle thro the window, and (like Dick Turpin, in the song,) "purvail on him to stop!"—shall have no respect for his age—dont care how *owl'd* he is!

I "laughed consumedly" at our Dutchman—who, as he tumbled in his bunk, remarked he "ish so glad it rains," for he "schleeps so sount, ven he hears to rain all night." The boys laafed too—they did, at the idea of hearing it rain when asleep; he insists he is correct, however, and this provokes a discussion, in which they drive him near distraction.

{Fred appeals to me, and I gravely decide it is a matter of fact, *he* says that he hears the rain when asleep, and the contestant says it cannot be so, evidently because *he* cant, which is irrelevant and inconclusive. and the only convincing thing would be for the disputant to lie awake some night and see whether *he* heard the rain when he was asleep and if he couldn't, it would go to show that Fred was more gifted that way than he, and anyway he ought to be the best judge of what he could and could not do, in the absence of any testimony to the contrary. and I turn back to my writing feeling that a jurist was diverted from a brilliant career when I took an ox whip in hand. but I do not feel sad about it.}[93]

......(Friday Morning.) Raining yet—and steadily, so that work is out of the question and we are all assembled in our little cabin again.

Oregon prognostics have entirely failed, this season. First a deep snow, which opened the eyes of the astonished "oldest inhabitant," who hadnt seen the like all his life, and which gave a chilly kid of welcome to the emigration of 1852. Then, instead of having March, as usual, wind up the Rainy Season with a "Grand Entree," it was mild and sunny—and now April, from which we expected better things, in a coquettish way smiles and frowns upon us; leaving May (two months later than usual,) to bring up the grass, and make flowers shine out and gem the land, like "stars in a field of emerald."

Oh! it is pleasant thus to sit listening to the gentle prattle of the rain, falling on the roof like the tread of angelic *infantry*—ever and anon, with pen arrested in air, and *revetren*tially pausing to look and listen! "This *is* to live."

The eagles which are building in the dead fir near the house, whistle shrilly as they soar circling overhead. We have granted them immunity now, for their neighborship—and tho occasionally the male bird alights upon the cedar in front of us, and not two hundred feet from the window where I write, (whence I shot the other,) we only look at him, as he sways to and fro in the storm, and molest him not.

The Indians have forsaken the bay to-day. Not a solitary aborigine or canoe is to be seen. Here goes the Custom House Boat, towing down a large ships-boat of something, we cannot see what. The rain is increasing, and (as I tell Ensign, who looks over my shoulders,) only needs one thing to render it, to sailors, perfectly aggravating—and that is, when completely wet, to have the sun come out and taunt them with their dampness! "Lo-in-b'hold!" out it comes as I speak—warm and bright! a southerly wind driving the rain down the Sound. The tall masts of the Cyclops[94] emerge from the mist. The crows sun themselves on the cedars. The damp oarsmen raise a little smudge of a sail—which is immediately blown overboard, a whole volley of indistinct curses float in shoreward, and the crows caw out a prodigious laugh at the disaster.

Our Dutchman has begun to talk energetically of going up to Olympia, the day promising to turn out so pleasantly—but the sun disapears again, the rains descend, and he "believes, on the whole, he won't go." Meanwhile I resume my letter. Ensign reading *Putnam's Mnthly*—another, in his bunk, singing vigorously the "House Carpen-*ter*," containing verses enough to keep him employed all day—while the Dutchman smokes phlegmatically his eagle-bone pipe, and remarks that, if he "on'y hat some goot *lager*" he would be happy. Fred., by the way, was born near Bairenth;[95] and, with the simple picturing of German life yet lingering in my remembrance, from the perusal of the life of Jean Paul Richter, I have been much interested in his description of his native town and neighboring country; his linings of the old farm-house in the "Faderland," where his honest old wooden-shoed parents yet reside, call up all the delightful pictures of German life, so well depicted in the volume alluded to, that I could almost fancy the "Fihtcobergs"[96] before me, with their quaint surroundings!

One of the many singular changes in my life and living, resulting from my exodus, is the almost entire absence of female society. I have spent some evenings at Olympia, in the hospitable dwellings of friends, but a visit to town involves a journey of half a dozen miles, enough to cool one's ardor. One day last winter when out hunting cattle on the prairies, I called on Judge Y—,[97] a hospitable and warm-hearted old Kentuckian, with two charming daughters. I essayed to sing a little, at their request, with great diffidence—and *eat* a great deal, with much less. I slept with the Judge—all of us, indeed, occupying the same room. I had not yet arrived at that state of apathy, somewhat necessary in a new country, which would permit of my disrobing before "company," and so (merely denuding myself of coats and boots,) "accoutred as I was" I "tumbled in." The Judge having no such scruples, divested himself of all and followed—disclosing, in the operation, "a mighty slim chance of legs." I have never chanced to pass the old judges since, but he is "a poplaur institution," and his daughters no less so; and at all times may be met parties of young men, "going out to judge Y—'s."

......It is night again. I am still writing—as I have at "stolen times" for a week or more past; which the unstudied, disconnected sentences and subjects will most certainly render perfectly manifest as you read. I listen, as I write, to the boys talking over their route across the plains—and the tears almost rise involuntarily as I hear tales of the many sorrows of emigrants, bringing back some of my own sad experience. The burials by the wayside—visions of tents by the road, where swaying of the canvas would disclose the dead or the dying—Weeping children—starving women—God-forsaken looking men—groups gathered around a broken down team or fagged out oxen! All these, and many other heart-rending sights, gliding by—a mournful panorama, almost making me live over again the desolation of heart experienced and endured when—crippled and worn down—I limped my weary way, from Ft. Boise to the Dalles.

Each has *his own* "ower true tale" to tell, and they talked about it quietly and subduedly—and, strange to say, they would be willing to travel it over again, provided *some* of the places were left out. I would gloriously like to undertake it with William!

We have got along finely with our contract—too finely for we have but about a week more to finish it. The men are all paid in full, now, and the original capital untouched. Must endeavor to get an equally good job, as we shall not remain long doing nothing—get too fat, else.

The mail is in, I know, to-day; arrived this afternoon—and there are letters for me, I am certain; never arrived a "States' Mail" without bringing one or more. None but an Oregonian or exile can appreciate the anxiety with which the monthly arrival is looked for, that "brings tidings of comfort and joy." I recollect, in my schoolboy days, (even then Oregon posessed a mysterious interest to me,) a picture in "Malta Brun's Geography," representing Oregon. Astoria was represented on the left, on an eminence, looking out upon the ocean and the breakers that guarded the entrance to that dangerous harbor. Three or four figures were grouped, peering anxiously thro a telescope to a ship in the distance, bringing them doubtless news from afar. It wasn't much of a picture—artistically speaking—for the designer was evidently of the aspiring class,

"Whose amazing style
Would make a bird's beak plain, at twenty mile;"[98]

—and had made sad work with his *chiaro obscuro* and foreshortening—the telescope almost reaching the ship—but it was very suggestive, and I could well appreciate the feelings of that pictured group, as they brought their glass "to bear on the ship" in anxious expectancy, to ascertain if she was American or not. With much the same feeling, the more certainty of its gratification, I watched the light gleaming on the little office at Olympia, from my window—three miles from me—knowing that, to-morrow, I will be with [y]ou all again!

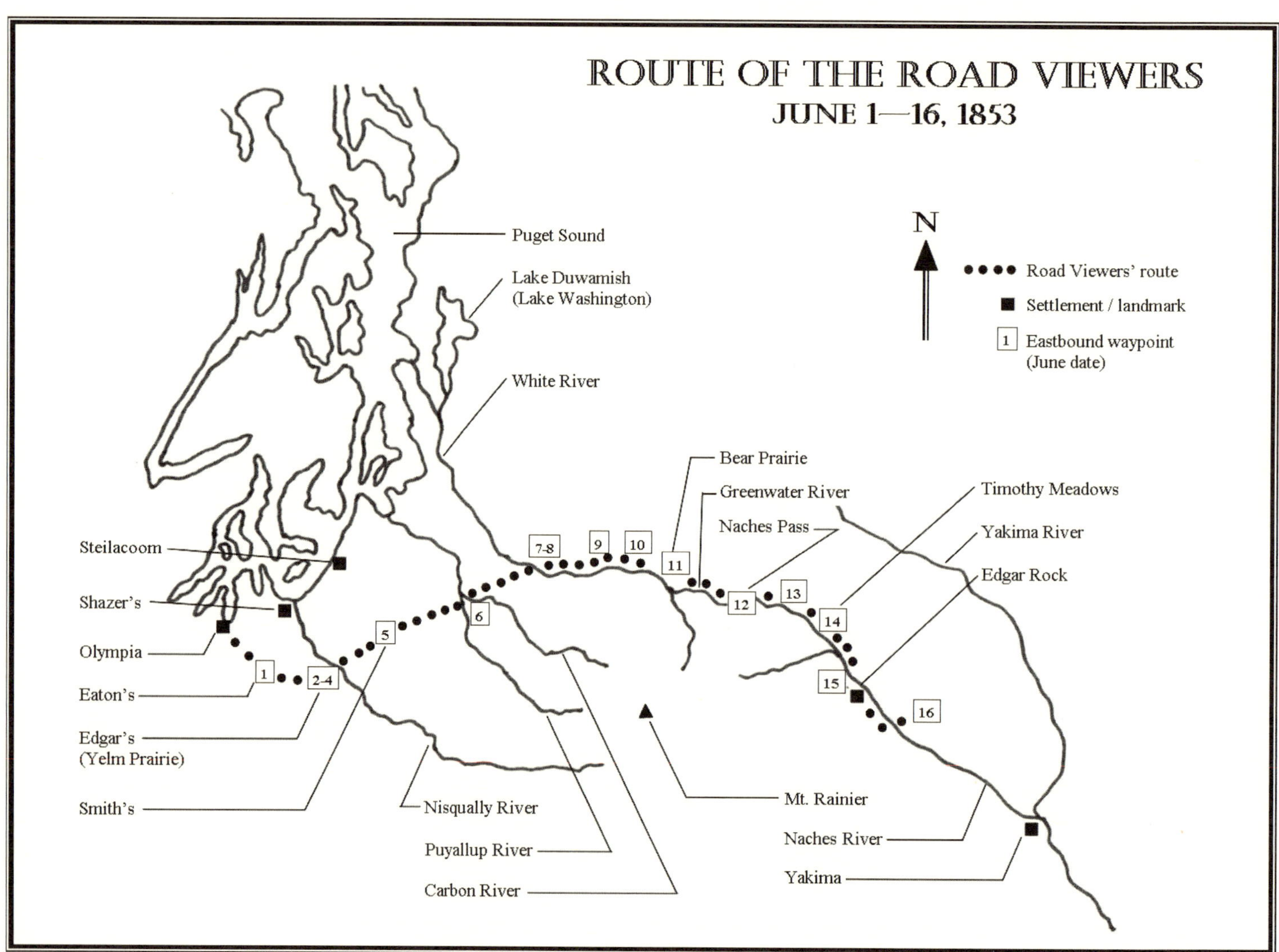

The day-by-day progress of the road viewers in June of 1853 is shown on this map of the Naches Pass area. Map by authors.

The Road Viewers

A Congressional Appropriation

In 1853, perhaps influenced by the efforts of the Whig party and others over the previous two decades to promote internal improvements such as roads, canals, and railroads, the U.S. Government took the first concrete step towards building a railroad line across the country. That year Secretary of War Jefferson Davis sent out four surveying parties whose purpose was to ascertain the best route for building a railroad from the Mississippi River to the Pacific Ocean. Leading the northern survey party was the newly appointed governor of Washington Territory, Isaac Stevens. Along with the news of the creation of Washington Territory and of the railroad survey came confirmation that $20,000 had been appropriated to build a military road from Puget Sound over the Cascade Mountains to Fort Walla Walla. The building of a railroad, everyone knew, was years away. But the prospects of a military road were palpable and exciting. Such a road, besides serving military contingents, would create a direct route into the Puget Sound country, allowing overland emigrants to avoid the much longer route down the Columbia River to Portland and up the Cowlitz Trail to Tumwater and Olympia. To say that the citizens of the new territory deemed such a road important would be a vast understatement. They saw it as vital to their future prosperity.

On April 30, 1853, the *Columbian* informed its readers that the money for the road would not be available until the fall, making it clear to the settlers of the new territory that no government road would be built in time for the arrival of wagons already working their way west that year. If a road was to be built, the settlers would have to act on their own, and quickly.

The editor urged his readers to take the road building matter into their own hands.

> The road will be wanted late in the summer, and we can do much towards it if we will only TRY...Let us all put our shoulders to the wheel, and amid a general hurrah, make one grand effort in this important matter. Late advices from the States say "an immense throng will cross the plains the present year"—and our word for it, many thousands will come to Washington, if they can get here...To the work, men! Time flies; we have none to lose...subscription papers are about being put in circulation...Give your name and your dollars. Now is the time to act. Let there be no more talk—too much has been lost by empty talking...and if to the question, shall we make the road? your answer be those words of power, 'WE WILL!'[1]

The following week the editor fantasized that if a road were built, some 10,000 immigrants would likely come over it that season. And a route already existed.

Early Northwest residents knew of a trail over the Cascade Mountains from Puget Sound over Naches Pass (4,920 feet) to the Yakima River country. Native Americans had used it for years as a trade route. The Hudson's Bay Company found it useful for the same purpose. In 1841, Lieutenant Robert E. Johnson of the Wilkes United States Exploring Expedition traveled over Naches Pass following the route of the old Indian trail, as did Father Francis Blanchet while conducting missionary work among the Native Americans (Father Blanchet was one of the first Catholic priests to come to the Oregon Territory. He served in the Northwest until 1844). In 1850, Michael T. Simmons of Tumwater and a crew of workers actually made an attempt to build a wagon road over the route. They cut six miles of road between the Puyallup River and Connell's Prairie before giving up, finding the task too daunting.[2]

Choosing the Road Viewers

On May 21, 1853, Olympia citizens held a general meeting to discuss the proposed wagon road.[3] W. W. Plumb was appointed chairman and A. W. Moore, secretary. William W. Plumb came from Michigan to Portland in 1849. He took up a donation land claim in Olympia where he served as justice of the peace; he also enlisted in the Washington Territorial Volunteers in 1856. Plumb eventually went back east, but returned to Thurston County and took up another homestead; he died in 1898. Andrew W. Moore arrived in Olympia in 1852 from Vermont. He taught in Olympia's first public school, at an annual salary of $25. He also opened Olympia's first drayage business, and served as Olympia city treasurer, Thurston County auditor, and clerk of the U.S. District Court. He died in Olympia in 1875.[4]

It was not at all certain at this time that the Naches horse trail, the assumed route, could be traversed by covered wagons. At the meeting, Edward Jay Allen read a statement furnished by "Mr. Blanchett" that gave information about the route and discussed its feasibility as a wagon road. George N. McConaha spoke of the advantages that such a road would give to the territory and "the necessity of every citizen lending aid to the furtherance of the project." McConaha was a Seattle attorney who had served as president of the "New Territory Convention" held at Monticello, November 1852.[5] He moved that a committee of five persons be appointed to view the proposed route and report on its practicability. Subsequently, McConaha, Whitfield Kirtley, Charles Eaton, John Edgar and Edward Jay Allen were chosen as road-viewers.[6] By June 1, however, George Shazer had replaced Eaton in the scouting party and McConaha had dropped out.

Support for the Road Viewers

A. W. Moore moved that a committee be appointed "to provide an outfit for the viewers, to receive their report—and if favorable, to receive the names of those who might volunteer to cut the road, and also receive contributions of money and provisions for those working." Reverend B. Close, A. W. Moore, E. Sylvester, Jas. Hurd, and John Alexander were appointed. The attendees raised $128 on the spot to support the scouts.

Several new characters in Allen's story are introduced here. Rev. Benjamin Close was a preacher of the Methodist Episcopal Church. He had previously been assigned to churches in Wisconsin and Minnesota, and was then assigned to northern Oregon. Just after the congregation exited the Olympia schoolhouse after Close's first sermon (December 26, 1852) the roof fell in under the weight of a four-foot snowfall. Close spent the first months of 1853 holding services in various buildings around town. Close served at Olympia until 1855, when he was transferred to the St. Helens-Multnomah, Oregon district.

Edmund Sylvester was born in Deer Isle, Maine, and came west in 1843. He spent some time in the Portland/Astoria area before heading north to Puget Sound. He is credited with establishing the city of Olympia (officially platting it in 1850), although it was first named Smithter, then Smithfield after Sylvester's partner, Levi Smith. Sylvester ran a hotel, and donated land for a mercantile store, city park, and Masonic temple, and ten acres for the capitol grounds. He went back east in 1854, and returned with a wife, Clara E. Pottle. He died in 1887.

James K. Hurd came to Olympia from Whitley County, Indiana. He ran the Olympia Bakery and Beef Market and served as the disbursing agent for the citizens' committee for road funds. In 1854 Hurd made two trips east over the Oregon Trail to ascertain conditions of the emigrant wagon trains and spread word to them of the new wagon road into Washington Territory. Hurd also served as a second lieutenant in the territorial volunteers during the Indian wars. He died October 22, 1857, after being attacked by a wild steer, leaving behind a wife and infant child. He was twenty-eight years old at the time of his death.

John Alexander arrived in Olympia in late November 1851 on the *Exact*, the ship that landed at Alki Point with a party of settlers who went on to found the city of Seattle. Alexander eventually moved to Whidbey Island, and was appointed as one of the first Island County commissioners. He also served as a judge for Island County (Penn's Cove district) and was appointed a member of the first Territorial Agricultural Society. He died in December 1858, leaving a widow and children.[7]

The story of the Naches Pass road survey has always been incomplete. Allen's letters and manuscript allow for the first time a comprehensive look at that June 1853 surveying expedition. No other description, beyond some brief accounts in contemporary newspapers, has ever been found. Allen also introduced the other members of his surveying team: John Edgar, George Shazer, and Whitfield Kirtley.

Allen's Pittsburgh Manuscript set the stage:

Twenty thousand dollars judiciously spent, would make, not a road worthy of the name, but a feasible, if difficult way, to get into the new territory, and one within reach of the emigrants, when they reached Walla Walla. The thing to do now was to make a reconnaissance and determine from an emigrants standpoint whether any of the passes through the mountains would permit such a road, and if found possible, to demonstrate it by opening it sufficiently to pass some wagons through, and so be able to secure the appropriation to further improve it...Of the Snoqualmie Pass we could learn nothing reliable, and it was deemed too far north. Cowlitz Pass was not considered.

At the meeting, which was well attended, it was decided to explore Nachess Pass of the Cascade Mountains, and five men, of whom I was one, were selected to go out and view a route for the proposed road.[8]

Allen's contemporary account began in the midst of a letter dated May 22, 1853. He had just canoed to Olympia from his land claim about three miles north of town.

(Monday noon.) I came up this morning and (dashing into the [Post] office) soon got my portion. *Such* a package! five letters, all the current magazines, "Harpers," "Littell,"—old "Knick,"[9] with its familiar blue cover, lots of papers, "cords" of *Dispatches*, treatises on agriculture and essays on gardening. The post master grinned, said I would have to charter an extra packet to bring them over; a "siwash" lounging near kindly volunteered to send his squaw to help me carry them down to my canoe—had barely time to glance over the letters and see you were all well, and ascertained even that much with great difficulty, as the lines soon began to run together as I read, my eyes filling somehow all the time.

I found the town full of people, assembled to hold a meeting and consult on the possibility of cutting a road to Wallah-wallah, from the Sound, to afford the emigration an easier, shorter, and less expensive road than I had come in the winter. I had thought a good deal on the subject, while chopping on the claim, and in intervals of leisure, rainy days, &c., and had written some little suggestive articles for our paper, calling attention to it, and showing the great advantages to the territory, if the road *could* be made.

Very little was known concerning it, and it was generally considered impracticable. An appropriation has been made, but never disbursed, as the attempt to perform the work has failed.[10] Priest Blarcehet [Blanchet] had traveled it years ago, and his journal, (and what information could be elicited from trappers, Hudsou [Hudson's] Bay Co's employees,) was all we had.

The meeting was held to day, (May 21st,)[11] and I have just left it—have been chosen one of five to go out and view the road. We will go on horseback, taking Indian guides—and, as it is an object to have it reviewed, and if possible *made*, in time for this year's emigration, we are to start soon—immediately. When asked *how* soon, I told them "as soon as I could go down home, arrange things a little, and read my letters." Next Wednesday we start,[12] will be absent about a month—so, if you are somewhat longer in receiving another letter, you may guess the reason.

—And now, with a month's employment before me, in high health, spirits and hope, I grasp my letters, hurry to the beach, and one half hour brings me *home* again.

"The shore! the shore!
The pebbly sand,
The birchen door,
The leafy land,
The carved canoe,
The "wickyup" fire,
The wavelet blue,
The pine tree's spire—
Ho, ho! ho, ho! I'm home again!
Nor dropped the plashing oar in vain."[13]

Fervently,
EDW. J. ALLEN[14]

Allen's next letter was written after the completion of the road survey.

ALLEN'S CLAIM, *near Olympia*, *Washington Territory*, July 20: My last, (written May 21st, but mailed some time in June, and which—if the return mails are as irregular as those "outward bound,"—you may not have received till now,) concluded with my successful visit to the Post office, and my appointment as one of the viewers, to ascertain the practicability of cutting the "Cascade Road."

I neglected to mention, in my exultation at the receipt of so many letters for myself, from friends in "the States," that I did not forget [Shirley] Ensign, who had not heard from *his* home for a long weary while. When seated, cosily, with mine all arrayed before me, I made mention of none for him—but his bitterly disaqointed look was too much for me—so I said I *believed* I had a little note, or something of the kind, for him. I never saw a man "brighten up" so, in my life; he read and re-read it—but at last said, with a puzzled expression, that it spoke of "another" that had been written. I remarked it was a most remarkable coincidence, their mentioning another letter, when I happened to have another for him, in my hat—and so gave it to him. After perusing it he laughed heartily at the manner in which I had doubled on him—but, in the height of his merriment I handed him a *third*. The "mirth then grew fast and furious," and I had work to convince him my hat contained no more. In the exuberance of his joy he insisted on my reading them—and I soon had the satisfaction of knowing that "Dad Smith" was still flourishing, "Champ Williams" had a fine son, "Ezra Beldin" was married—and various other little scraps of domestic news, of an *anticipatory* nature, which I rather calc'tate were penned with no idea that any but the eye of a beloved father would ever glance over them.

A day before that fixed for sta[r]ting upon our expedition, I was sent for from Olympia, to come up and prepaire the necessary outfit, mules provisions, &c., to collect the necessary funds for providing all which another committe of five had been appointed.

Our own committee consisted of E. W. M'Conahy, who (from some unexpected circumstance,) was prevented attending the expedition—a lawyer of ability, and formerly of the California Legislature; urbane courteous, in politics a vehement Democrat, or a vehement anything else that happens to be in the majority.

John Edgar,[15] for fifteen years a resident of Oregon, once an employee of the Hudson Bay Company, and now an American citizen, married to an Indian wife and bringing up his family in a Christian-like manner—an honest upright man. Whitfield Kirtley,[16] who has been in Oregon ten years, has mined successfully, and is now established at Olympia, in the prospects of which he has great hopes—owns a good many lots ther[e], lives on his means, "doing odd jobs," occasionally—getting out timber, &c., as we did; full of energy and enterprise. George Shazer,[17] an original, and a comical kind of genius; an old residenter [resident] and mountaineer, who had formerley traded and trapped for years on the head waters of the Mississippi, making Fort Smelling[18] his headquarters—and subsequently accompanied Frémont in his explorations of that country;[19] he was generous to a fault, impetuous and persevering. Lastly, (and *least* in at least a physical point of view,) Edward J. Allen, of whom modesty forbids more than a mention of his name.[20]

The Ride to John Edgar's (June 1-2)

Our rendezvous was fixed at Edgar's residence twenty miles out on the prairies—each one mounted, with an extra mule or two, on which to pack provisions for a month. We started out, but finding one of our mules missing, sent a man to search for it on Brush Prairie in vain—and, concluding it had strayed on the prairie we were compelled to cross, determined to go on, and search for it as we went along.

We rode out South-west from Olympia, and after a ride of three miles thro dense woods came upon the open prairie—"Chambers' Prairie."[21] The saddle I had depended on for the occasion was a mile or two distant from the prairie, and I had to make a detour to get it—Kirtley, at the same time scouring in an opposite direction, to find the vagrant mule. There was an ox team going out twenty miles, part of our way, belonging to a settler named Eaton;[22] and, as the driver volunteered to take my horse, and help to search for the stray animal, I [t]ook the goad, and jogged on with the team, feeling almost like "an emigrant" again.

"Chambers' Prairie" is of small extent, but very beautiful—covered with grass, hemmed in by grand old woods. Far beyond us loomed up the snow capped Mount Rainer, seeming only at the edge of the gras[s] level.

We arrived at Eaton's[23] a little before dark—a cosy little log house, nestled in one corner of the prairie, the farm being in the bottom land below, and entirely hidden from view. Before leaving the woods, we passed a curious little lake,[24] about a quarter of a mile in diameter, far down in a deep valley, seemingly without inlet or outlet.

We slept at Eaton's that night, and (having recovered our mule,) made final arrangements for packing, early the next morning resuming our journey. It was a curious sight, to see Eaton's cattle coming up "to be salted." He had some two or three hundred, chiefly Spanish, naturally almost as wild as deer; they would come up and gallop like a herd of horses, then turn and scour over the plains, repeating the evolution perhaps a dozen times ere they finally partook of the salt provided them. Very wary were they—at the least alarm tramping off, in a body, two or three miles ere they stopped. They have been suffered to run thus wild, without shelter or feed, and have increased most wonderfully.[25]

A Few Days at Yelm Prairie (June 2-4)

We rode, that morning, over the prairie—and passing thro a narrow dividing belt of timber, entered "Yellow Prairie,"[26] which has a light gravelly soil, similar to the one crossed on the day before; and arrived at Edgar's about noon.[27] He was out in the fields, but came in as he saw us riding up. His Indian wife, (a good looking woman for a squaw, only they generally have such queer shaped heads, and such an ungraceful walk,) bustled around and soon got us some dinner. He seems very proud of his children, unattractive as their dull, almost African featuers are. While we were conversing, there was a great outcry among the chickens; on looking out, we saw a crow flying away with one. I was unaware, till then, of it being an unusual circumstance.

We found Edgar was not ready—nor would he be until Sunday, so we had to quietly settle down and wait for him. The Indians had told us we would find it impossible to cross the mountains, on account of the snow—that their annual exodus over them had not yet been made—and this somewhat reconciled us to [t]he delay.

Assembled around the fire, that evening, I heard many tales concerning the early settlement of Oregon, skirmishes with the Indians down about Shasta, traits of old Joe Lane, &c. &c. I had often heard that there was gold somewhere in the Yakima Valley, east of the mountains, and asked Edgar concerning it, he said there certainly was—as Indians had repeatedly offered it for sale, in Oregon City, but would tell no white man the locality where it was obtained; that the great chief of all that country was "Owohi"[28]—to whose niece he was married, and who had often entreated him to come and live among them, promising to show him where the gold was, if he would remain with them entirely. The "Tamananus," or "medicine man," has fallen out with the old chief, and this spring he was at Edgar's, and wanted him badly to go. Edgar refused, as he could not leave his crops, but gave him a small buckskin sack, and told him that, if he would bring it back full of gold, as an evidence that there *was* gold, he would return with him. He is now on this errand.

In the evening rode up a party of drunken Canadians, with their Indian wives—who rode in the same cavalier fashion as their husbands; and, from their expertness and skill, one would fancy had been almost born on horseback....

Last season, Edgar made a trip to Mount Rainer,[29] the view from which (as he describes it) must have been very beautiful, tho he did not go beyond the snow, which was thirty feet deep[.] He related to me many curious tales of the manners and customs of the tribe of Indians which "thereabout do dwell," a great deal of which is embodied in Wilkes' Report. They have generally very pretty names for the prairies—"yelm" "Fanalquoit," [Tenalquot] &c., as also for their rivers, lakes and squaws.

His place is a very beautiful one. Upon it is one of the best springs[30] that can be found in all the territory, gushing out almost large enough to turn a mill, and as cold as ice. He has a great deal of land in cultivation, and has had the good taste to leave standing, in one of his fields, a number of scotch firs, beautiful shade trees, and evergreens. In another is an Indian burying-ground. He is also very proud of his sheep, of which he has great numbers—regular "South Downs," and which were the same drove that preceded me in my journey from Portland, over the "Pack trail," to Puget Sound.[31] One of the drivers, Fammaree the Sandwich Islander, was here at Edgar's, and seemed borne down by the idea that I should recollect him, and evince it by speaking to him of it. He had lost a great many of them, coming over—fortunate to have saved any at all, indeed...

Talking to Edgar about the birds of Oregon led to my discovering him to be an Englishman, from the enthusiasm with which he talked about the lark, linnet, thrush, and other birds of the British Isles, the mention of which seemed to call up in his mind all his boyish days, and increase his determination to revisit the scenes of his youth.

This afternoon, after supper, after saying he would be ready to go in a day, he seemed low spirited—remarked he was on a long trip, where changes might occur, and was concerned to think that, in case of his death, only two of his children could inherit his property, the others having been born before his marriage. Having, at his desire, prepared a rough draft of a will,[32] giving to his wife a life interest in his estate, during widowhood, and providing that all his children should inherit the property, equally, after her death or second marriage, he took it to Nisqually, signed it, had it recorded, and left the original in the hands of Dr. Tohuie. [Tolmie][33]

This being "States Mail" day, and Mr. Edgar having to send an Indian to town, I wrote a brief note to our postmaster to send anything for me out. The Indian returned with a welcome bundle of letters, &c., and a neatly done up package of twenty-five varieties of seeds of vegetables and fruits, with appropriate and specific directions—which must have cost a great deal, coming by mail. Edgar was very thankful for some of them.

Journey to the High Hills of Puyallup (June 5-7)

Next day we were all ready; and started off—a cavalcade of fifteen or eighteen horses, at last fully on our way. Leaving Yellow Prairie behind, and crossing the Nisqually river, (a little stream, about fifty yards wide and not very deep,) we came out on Long Prairie, and passed thro Hell Gate, the sulphurous name given to a gap where the bluffs approach and form a narrow

This engraving of the Puyallup River from *The West Shore* magazine of July 1883 depicts the terrain the road viewers had to navigate.

opening to the next prairie, known as Muck—crossing which, and passing a settler's claim, (Sandy Smiths),[34] the last house we will see henceforth, we camped on Muck River, under a wide spreading tree. Here commenced our mountain life, and we cooked some supper, while others washed the mud off the animals' backs before they had time to cool, to prevent them becoming sore, and staked them out on the grass.

In the morning early, having long lariats to the horses, we turned them loose, expecting little trouble in catching them—nor had we, with any but mine, which was "a tarter," and wore us all down in the endeavor. Having crawled on my stomach a long distance, concealed in the high grass, and succeeded unobserved in getting hold of the trailing lariat—it was only to find that 147 lbs. at its end was a trifle to him; I must have presented a comical sight, snaking thro the tall grass at the rate of a mile a minute, directly in his track, but gaining never an inch—and, approaching rougher ground, where sliding might prove less agreeable, was glad to have the assistance of Shearer [Shazer] and Edgar. Once saddled and mounted, he was an excellent horse—full of fire and deviltry, tho.

Getting under way once more, we forded Muck River, emerged on Island Prairie, which we crossed to the bordering woods, passing thro them to Silquito [Selquite] Prairie; another belt of woods, and Kado Prairie; ditto, and Echo Prairie[35]—entering which we suddenly came upon a herd of wild cattle, (belonging to the Hudson Bay Company,) which had been running wild so many generations that they were as timid as buffalo. *Such* a scampering! "Heads I win—tails you loose!" and

> "Away, like the wind, pausing not to look back,
> For the shedder of blood was quick on their track!"[36]

All were excited in the pursuit—and plunging in the midst of the herd, some were about to fire their revolvers, when a timely word prevented a slaughter, which must have been most wanton, for we lacked not provisioni [provisions]—some of the Indians coming to our encampment with more dried salmon and delicious berries, in such quantities that we had to leave a quart of the latter untouched.

After supper, we mounted again; and, thro a few miles of timber, came in view of Pur-la-lop Valley—very deep, but down which we formed a pretty easy descent. There we delayed some time, marking out the best place for our new road—and, this accomplished, entered the Valley.

Crossing four different branches of the Chewah River,[37] (at the last of which was a bed of quicksand, into which an old horse, with a yet older squaw on his back, "put his foot"—and such plunging I never saw executed before!) on the other side we rode thro a salmon-berry patch nearly a mile in extent. There was a good deal of what is called "vine maple" growing thro it, which has a leaf nearly resembling our water maple—but the trunk rarely exceeds in growth the size of "a thin man's leg." These overhung the trail so much that we had to bow down on our horses' necks, half the time, to pass under. The Indians enjoy a quiet joke—but it must be practical, or they cannot understand it. They pretended to be very diffident and retiring, deferring much to Shazer; I, naturally of a modest disposition, also held back—so that Shazer was our leader; it had been raining slightly, and the first one to pass thro the thicket would necessarily stand the brunt. It was some little time before he thoroughly understood our unwonted modesty.

We camped on Mamolamee[38] Prairie, (which signifies "many-rooted,") covered with many varieties of grass, among them a beautiful scented clover—

"And what a wilderness of flowers!
It seemed as tho, from all the bowers
And fairest fields of all the year,
The mingled spoil was gathered there!"[39]

Crossed Twallep [Puyallup] River the next day! very swift and shallow—the stream being about fifty yards wide. It heads (I think, from its course,) in Mount Rainer. Half a mile further we crossed another beautiful little stream[40] across which the Indians had made a salmon dam, similar to those described in former letters; there was but one old Indian fishing, (all the others, having gone over the mountains,) and he told us doleful tales of the small-pox being very bad across the country—many dying, a result which almost invariably occurs, as when attacked they immediately plunge in cold water, which appears to kill them almost at once. During the prevalence of disease their "medicine men" are in great danger, for the increase of mortality and failure of their incantations, is frequently punished by their death—rather a severe test of their skill and judgment.

At the high hills of Puleullup[41] (well named,) and up which the trail lead, we halted for an examination, and finding a wagon road there impracticable, detached a party to explore the base, who found, to the north-east, a course by which the difficulty could be avoided.[42] After "blazing" the route we returned and resumed the ascent of the hills, having a hard time in doing so, compelled to drive our pack animals before and lead the others—keeping meanwhile a sharp look-out to avoid anything rolling back; and well I did so. We had in the train a huge stout Indian horse, which had a trick of lying down on the march, with his load or his rider, and remaining in that interesting and provoking position till it suited his dignity to rise; he had done this frequently, during the day, to our great annoyance and detention; I watched the look he gave his leader, as he jogged along, and could imagine I saw that in his eye which boded no good—and that, in his opinion, there could be no better time nor more opportune place to play his old game; no sooner thought than done—he sinks back on his flanks, quietly down, but had mistaken the ground and turning head over back executed a series of "ground [grand] and lofty somersaults," with his well-secured pack, surging past me, (as I kept clear of his legs in his descent,) rolling over and over, till he "came too" on a little kind of a bench, about thirty feet below—the most astonished animal I had yet seen! We had no trouble with his tricks subsequently; that adventure cured him effectually.

A short distance from here, our Indian guide, (a stalwart old hunter, honest man, and one of the finest looking specimens of the aborigine.) left us. All along, while we were riding, he had walked, stalking on in advance of us, and never losing sight of or deviating from the trial [trail], however indistinct at times it seemed to us. He had only designed going to White River, to procure a peculiar herb and root that grew there, and then returning to his wigwam by a "short cut." Our occasion to regret his departure will hereafter be alluded to.

The view from here was most beautiful, surmounted by alternate prairies dotted with 'motts'[43] of trees, like oases, and fringes of woodland, beyond which the eye swept over an interminable forest, me[l]ting into a blue haze from which Mount Rainer, capped with eternal snows, rose to a height of 12,330 feet. The purpelness of the atmosphere is a feature in the landscape, I think peculiar to this country. We were twenty five miles distance from Mount Rainer, tho apparently only one fifth that distance.[44]

The descent of one of these hills, and the passage thro another belt of timber, brought us again upon one of these little prairies,[45] covered with gigantic pines of towering height, and branches drooping to the ground, varied by clumps of elder, maple and oak—covered with the various species of grasses which spring up luxuriantly in the spring, and (ripening in July,) form the natural hay cattle prefer to anything else. The grass is so emeraldish green, flowers so numerous, trees grow so regular and graceful, that it is difficult to conceive that nature above has developed so much beauty. We dashed over it at a full gallop, all eager as much

"To pass
The virgin verdure of the wilderness."[46]

as to obtain some water from an Indian camp, which we descried across the prairie—and which was generously given, as also some venison. They were going to cross the mountains, but had halted, from rumors of the small-pox being prevalent. Here is the place where the old road intersects, for this is the second time it has been projected. There was an amount subscribed before to cut it, but the contractors never viewed out a new road, but merely followed the old Indian war-path, leading from Nisqually.[47]

It was only a mile from White River or "Tcope Chuck," appropriately named, as all Indian nomenclature generally is—the water being of a milky white. It was pretty high, so that it was impossible to ford it. We had to unpack and unsaddle our horses, and drive them across first and then, with the assistance of the Indians, pack our goods across a kind of bridge formed by the stoppage of the sage trees, and accumulated "debris," brought down doubtless by the spring freshet. The river was 200 yards wide, and remarkably swift. We all, however, succeeded in crossing safely, and worked up to a bar a mile above, where we camped, and drove our stock to a little prairie further on.[48]

There was a pleasant interchange of thoughts and confidences that evening, as we were assembled around the campfire, the light of which flashed out on the night, illuminating the dark woods and irradiating the waters murmuring by. A motley group were we, assembled there; the stoic of the woods disposed in picturesque figures around—and many and various were the experiences related and opinions avowed;

"And freely spake each heart
In its native tongue—the wisdom taught
At that wondrous school of life and thought,
Wherein men learn apart:
And which came nearer to the way
Of the stern old truth—let sages say,
If they e're take note of the poet's lay."[49]

Having lost the assistance of our Indian guide, it was necessary we should procure another; we therefore remained there all next day, awaiting the coming of another Indian, whom Edgar, (whose marriage to the chief's neice gives him great influence,) had sent for. We had, so far, been traveling a north east course from Olympia and were making good progress—as, in spite of all detentions in halting to mark out various routes we explored, we had come ninety miles. Wilkes' description of this country will apply pretty well to this portion of the route, as he traveled it in June 1841, a little northwardly of us, thus far on his way to Forts Okanadagua [Okanogan] and Colville. Edgar tells me that his Indian kin who occupy, six months in a year, this country on the west side White river, say that there are prairies just like the ones we have passed thro with a richer soil, composed of a clayey loam, with a vegetable mould over it—the prevalent trees being poplar, maple, ash, fir, pine and cedar, with laurel all growing to an immense size. Some of these prairies and lowlands are overflowed with the spring's freshet.[50]

Incidents Along the Trail (June 7-10)

While camped there, an Indian brought in a "dear little deer," he had picked up on the prairie; it was very young and graceful; fawns at that age are fearless and confiding, and will follow any one around. I urged the finder to return it where he got it, that the doe might recover it—but an Indian cannot understand such feelings, and so he left it with one of our tawny followers, and went his way. The poor thing went bleating around the camp, till I could not stand it, and fell considerably in Indian estimation as "a young brave," by feeding it with berries and bread.

If there is anything in the world that excites me and "makes my angry passions rise," it is cruelty to a dumb animal. I have never recovered from my hatred to New Orleans, occasioned by a first sight of the atrocious inhumanity to brutes in that city, where draying is a powerful interest. In consequence of the transfer and reshipment of the immense produce poured down from millions of acres, it is customary for them to buy their mules to "use up," and buy more every two years—and their appearance and treatment is enough to make one's heart ache.[51]

I have been surprised that no tourist has taken note of this, and had enough of the Englishman in me to be pained by it, and get involved in continual "musses" in consequence.

My trip over the plains, and daily contact with that noble animal, the ox, has developed this repugnance to all such cruelty. No greater compliment, in my opinion, can be paid to one, than to say "he has a heart like an ox." What does it not imply? Courage; calm; resolute endurance; indomitable perseverance; patience, generosity and impassive equanimity under all difficulties.

But, to return to the little episode of the fawn. One of the Indians came, to kill it, laid it down, and (while the creature was licking his hand) commenced sawing across its glossy throat with a dull knife—so dull, in fact, that he had to cease, and go in search of a sharper implement—the fawn, meanwhile, instead of endeavoring to escape, following him about till it dropped dead! It was best to kill it, as it was so young that, left to itself, it must have died from starvation—but I felt impelled to "fetch him a kick," for the bungling and heartless manner of doing it.

It rained all next day, and as we were unwilling to proceed without our guide who had not yet come, we remained in camp all day, making a comfortable cosy little tent with our horse blankets. I got Shazer (as usual on every opportunity,) to relate, in common with the others, some more of his "life yarns." He has spent many years among the Sioux and Chippewa Indians, and asserts they eat portions of the flesh of their enemies taken in battle. The Sioux once invited him to a grand feast, and regaled him most sumptuously on dried bear-meat—telling him afterwards it was not *all* bear-meat, but a compost, a portion of it being "dried Indian." He remarked, very gravely, that it "sot him agin bar-meat ever since."

Edgar's turn coming next, he "norrated," what wild times they had on Sophy's Island, in the Columbia, below the Willamette River, hunting the wild cattle and hogs, which had grown so numerous and destructive that it became necessary to wage a war of extermination against them—and also an account of the eruption of Mt. St. Helens, sometime in 1842, which threw ashes as far as Vancouver, and covered the pasturage there with a kind of sulphur, that killed off the cattle so fast they were compelled to move them to Nisqually.

The rain continued all night, and when morning came it was *still* raining; when, finding our Indian not come yet, we rolled out, and (ascending a small hill,) came into a beautiful claim of small prairies and small timber, seemingly extending as far as Sneckonish River,[52] probably 33 miles, beautiful rich land, luxuriant with flowers, red and white honeysuckles, and every variety of berry in bloom. The sward in some places was smooth and regular as an English lawn, while the trees, grouped here and there as if by art, made it seem as if an English Nobleman's estate had been swooped up bodily and dropped in this mysterious land.

Here, again, we examined the ground and localities, to see where our road should go, to avoid the high hills; and having found a route to our satisfaction, resumed our course again. Descending into White River Valley, we followed along the bed of the River, until we came to where it narrowed so abruptly as to compel us to ascend a tremendous mountain,[53] and along a ledge overlooking the valley, the river itself flowing 600 feet below, tumbling and foaming down the kanyon, from the mountains, reminding me of the scenery of Lewis' River.[54]

The next trouble we had was the coming to the first of a succession of deep valleys and hills, where we had a tight little job of unpacking our mules and horses, packing the loads ourselves, and driving the animals before us along the narrow ledge. On the top of these hills the trail led up a kind of little hollow; where, to pass under the trees which had blown across the road, we had again, in many cases, to unpack and make up in smaller bundles and oftener carry ourselves. These forests were of spruce, and measured (those that had fallen,) 268 feet long, and 35 feet in circumference, many of them being clear of branches for 150 feet! They stood so high it was impossible to see over and beyond them, even on horse-back. Seedlings and sprouts had sprung from the decaying bark and trunks of these fallen veterans, reaching the ground, and looking not unlike the Banyan tree, forming secluded bowers and cool sequestered arbors, through the trellis-like leaves of which

> 'at stolen whiles,
> Glinted the stray sunbeam or the meek azure smiled.'[55]

{Shazers horse was evidently not accustomed to mountain trails and balked at the tree scrambling. Shazers idea of instruction seemed to be to spur the horse to jump, and while he was in the midst of his scramble, to fire as many chambers of his revolver as could be exploded before he got completely over; it was an original method and it certainly incited the horse to speed for he nearly went frantic in his haste. Instead of balking, when he came in sight of a tree across the trail, he did not wait for the spur, but pitched forward headlong to make the effort. You might suppose he would endeavor to throw his rider, but he was an Indian horse, and knew the moment he felt the pressure of his riders knees what manner of man was bestride him.

{Had I been on him, I do not think he could have thrown me, but most assuredly he would have made a vigorous try for it, and that is the difference between my seat in the saddle and Shazers and the horse would know it. We *ride*, but fellows like Shazer are part of the horse.}[56]

Night was settling down on the grand old hills, and leaving behind us the murmuring of White River, we went over half-mile and camped by a little spring of pure water.[57] The ground was tolerably damp, but that is no drawback to a mountaineer, and after a supper of hard bread and ham, we tumbled ourselves in our blankets, by the blaze of a good lodge fire, sleeping right across the trail, to prevent the cattle from going back. We were wakened in the night by the noise of their tramping by; knew they could not go far, so took it easy, and in the morning they were all brought in early, by the Indians.

We travelled a long distance next day, before reaching White River again, and followed it occasionally, leaving it at intervals to cross over little spurs of the mountains, covered with pine-trees of gigantic dimensions, and returning to it again. About sunset, we were compelled to come down a nasty little hill,[58] where we had much difficulty in preventing our horses from tumbling in the River, flowing below. A mare of Edgar's, heavily packed, *did* get in; the current was running like a mill-race, and swirling and roaring to such an extent we could hardly hear each other. Her little colt was standing on the bank, ready but fearing to jump in after. I kept hallooing to Edgar to bring the lariat he had in his hand, and secure it, before it would—but the roaring of the waters prevented him hearing me, and in it went! I thought they were both lost, now, but the mare, heavily loaded as she was, managed to make the bar below, and missing her colt, walked up the bar and made out to swim across.

Edgar is very expert with the lasso, and guessing the mares intentions, I sang out for him to lasso her as she swung past, and float her gently in the eddy below. But he was afraid of strangling her, so she waded out to the deep water, but the moment the current struck her, she was swept off her feet, and whirled past the bar[.] She regained herself soon, swam nobly, but could not make the shore and boomed by where we stood—till her pack caught on a snag in the channel, which arrested her course. The water was surging and boiling around her overwhelming her till for half a moment her head was out of sight, and Edgar said it was all over[.] It came up again, and the poor animal snorted and looking so pitifuly, for help, we could not stand it I threw off my boots and coat, took the lariot,

The Indian Trail followed this section of the White River west of Naches Pass. Photograph by Dennis Larsen.

and a small rope and swam almost down stream to the snag. I had calculated to a nicety, and just hit it; got on the snag, (which swaying to and fro,) tied the mare with the lariat *to* the snag, and the pack to the smallest rope, then cut it loose from her back, and swung it safely below. Then bracing myself against the snag, with my feet to the shivering animal, shoved her out in the channel, when she landed softly at a point below, to her great relief, and mine—glad to find myself once more upon "Terra Firma," and the swirl of the waters no longer in my ears.

This little incident incurred to me the fervent gratitude of Edgar, with whom the mare was a special weakness; and reinstated, in the minds of our Indian allies, their former good opinion as to my being a "brave," which had been somewhat lessend by my weakness in the fawn affair. I was considered now more as a "Tice."[59]

The Climb to Naches Pass (June 10-12)

I was awakened, a little after day break, by a great shouting, and looking up, found our Indian guides had come. They had found our campfires on White River still smoking, and had ridden hard to overtake us—so we all rode nimbly off together. One of them was an old man called "Tcope Letebe," or "the White Head," the other was a monkeyfied little fellow, with a droll kind of a skin cap on, that ended in a tail hanging down behind—he riding a splinterlegged poney, and the old man a powerful, large black horse, with untrimed hoops, [hooves] that looked as ragged as a bunch of triangles.

We rode past "Letebe,"[60] or "the Head"—a singular looking peak or headland, coming boldly up to the river, devoid of timber, it having all been burnt off. I opine, from the top there would be a most extensive and beautiful view; but the haziness and unpleasantness of the weather and our lack of time, compelled me to forego it. We crossed a beautiful little stream, called Greenwater, which empties into White River just at the crossing {of the trail}. Here we cut across a bend, and left White River entirely; and at night encamped on "Ichfoot" or Bear Prairie;[61] and here Shazer went out after a bear—and here he did'nt find any! In the morning, we left the little Prairie, commenced the ascent of the main ridge of the mountian, and if you want to have an adequate idea of the dangers and difficulties we encountered in so doing, read the account by Wilkes who crossed them ten years ago, at a point considerably northward of us. He detached a party to survey Letebe, the promonotory spoken of above, and found it to be, by barometer, 2,793 feet in height, destitute then of wood, and the fires still burning, which had left it entirely bare when we passed it.

Hitherto our route had led us thro canyons, valleys, around and over hills, and along ledges; but at this place the trail led right to the foot of the mountain. After riding a few miles we passed a low swampy place, at the right of the trail, and here the Indians halted, and cut a small bundle of grass for each horse. I wondred somewhat at this manouver; but Edgar toled me it was a custom of theirs—that it was considered by them unsafe to omit the ceremony; he also urged us earnestly to say little, as we passed—"Hiyon wawa-hiyon-snass," (much talk, much rain.) This they religiously believe; and also that much talk would bring snows.

An Indian marks out a trail in a curious way; being generally destitute of an ax or any implement to clear out the sap[l]ings growing in the way they always endeavor to wind around where the timber is scant, and either climb over or go a round any fallen timber. They have no guide or compass, save eyesight. They make a course for the first tall peak, and, ascending it, look for the next one in line; and, ascending that, course for the next one—and so on. Thus we crossed up on high knolls, to go down the other side, only to illustrate a juvenile distich:

"Jack and Jill
Went up the hill,
And then—went down again."

An Indian hardly ever leaves or alters an old beaten trial. It has only been about fifteen years, I think, since the father of old "Oachi," [Owhi] whose country is in the vicinity of the Yakima valley, on the east side of the mountains took a notion in his old flat head to cross over to the west side of them, see the end of the world, and ascertain how things in general were going on there. He came over the Nisqually by this trail, which he was the first to open, and took back with him a number of the Nisqually tribe, as slaves—a disastrous speculation, that he repented for years after, and be hanged to him!

We crossed Greenwater many times during the morning! and, at last leaving it, came to the foot of the mountain and commenced its ascent, a wearisome and difficult task; a weird-like gloomy looking place, made doubly so, by the sombreness of the day, which was cold and hazy.

"Where the huge pine reddens
The rocks and soil with its rusted leaves
And skeleton cones,
And the footstep deadens,
As we clambered o'er roots and broken stones,
Whilst a noise of waves the ear deceives,
And the sigh of the wind, thro falling leaves,
And the restless heart saddens
With the surging tones."[62]

One old mule whisked his big head indignantly around, and (giving us a contemptuous look,) turned down again—but a few lashes with a rawhide, and some vigorous punches, corrected his little error, and he achieved that ascent with unusual vigor. "Essex," my horse, tramped up, after me, (with the punishment given the mule in his mind,) so energetically that I had to "hump it" quick, to prevent being run over.[63]

Arrived at the summit, we halted and unanimously indulged in a long breath. Our "Indian allies" began to strip the bark off the pine trees near, and scraping the sap of the inner bark began to eat it. It was a tree called the "sugar pine," with sap as sweet as that of the sugar maple, and which (the Indians tell me,) in the spring exudes a mass of sweet gum, not unlike congealed honey. Wonder if, some day, it may not be to our territory, what the cane and maple are to "the States."

Camped upon the mountain prairie,[64] (Lamuti Klosh Illiha,)[65] one side of which was bare, and the northern side (tho but 200 yards wide,) covered with six feet snow!

It had been raining all day, as I told you—and a miserable kind of a drizzle, worse than it was below—a kind of determined sleet. Our hard bread was all used up, and we had to bake some. We had a little pan to mix some flour in, and browning it a little in the frying-pan, before the fire, to such a consistency as enabled us to handle it stood it up on edge, sustained by some twigs, set in the ground, and baked it thro finely—and quickly, too; an impromptu mode of baking!

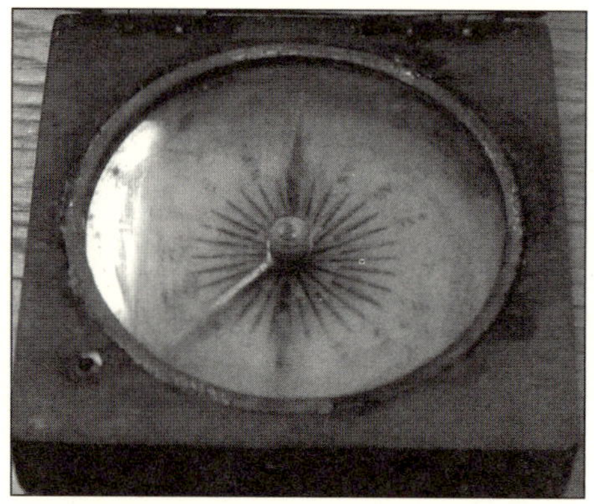

Allen described looking at his "little compass" on June 12, 1853, as the road viewers were descending east from Naches Pass. The compass is now in the Hervey Allen collection at the Hillman Library, University of Pittsburgh. Photograph by Dennis Larsen.

We were encamped under a nice, compactly leaved and wide-spreading tree, which sheltered us from the rain, in some measure; and after sitting awhile around our fire, watching the shadows in the tall trees around, and drinking great quantities of tea; in which we had boiled some roots called wild ginger, very hot—we slid into bed, all wet as we were; and awoke next morning—dry, all but our faces, on which the rain had beaten, and our blankets, from which the moisture streamed as we got up.

Lost in the Snow (June 12-14)

Got some water from a small stream flowing under a bed of snow, and started off buoyant very early, that we might cross the snow while it remained hard. Here we were again cautioned by the Indians not to speak a word—and to ascend along a ridge, doggedly, thro the rain, in "solemn silence." Once upon the summit, to my surprise we emerged on a chain of small prairies, alternated by belts of timber. On some of these prairies we could see, thro the opening of the trees, snow—in some places ten feet deep. It was tolerably hard crusted, and we passed over, leading our horses, with little difficulty—tho they broke thro occasionally, and floundered in pretty deep.

Where we entered the woods, and began to descend gradually, I looked at my little compass and found we were travelling due south.[66] This I knew must be wrong, if we continued long, and finding we did, I mentioned it to Edgar, who questioned our guide, who confessed at last that he was completely bewildered.

Here was a nice situation for "a small tea party!" somehow, in that dull drizzling rain, and with the dreary prospect of snow all around us I could hardly imagine but we had come thro the summer months, and just emerged into winter. It seemed so thoroughly and completely winter, that I was as confused as our guide, tho my perplexity was caused by the inexplicable confusion of season—his from losing all trace of the road.

Our guide was the little droll-capped Indian lad. "White-head" had gone on before us. Edgar, taking a course from my compass, struck across, and, seeing the old man's bushy-footed horse had gone on too, we all struck his trail, finding it correct by the compass, and soon came to another eminence, upon the eastern side overlooking a small valley, where we found our old white-headed guide, who had grown mystified too, and did not know how we could find a passage down.

Things began to "look blue." Two points being given, to go from one to the other, I was for giving our guides, who had proved themselves inefficient, the go-bye, letting them find their own way down, and we ours. We had proceeded a mile, when they called to us, shouting they had found it—so we clambered up the snow again, leading our fagged animals, only to find they were mistaken, and as far from the route as ever.

Through the mist loomed up a peculiar sugar-loaf shaped mountain or cone, which Edgar recognised, and thro his advice we struck a course, found the trail, descended thro the snows and into summer scenery again, and arrived at the small stream in the valley,[67] where we immediately encamped, pretty well tired with our wanderings. We sent our horses to a prairie further up the valley, where grass was plenty, and stationed a guard over them.

The Indians dug up a great quantity of roots[68] on this, (called Ahnepash) prairie; they were much smaller than a potato, about the size of a damson plum, and resembled the artichoke—These they cooked in regular Indian style—digging a hole in the ground, filling it with hot stones, covering them with grass and fern,—sprinkling water on the grass, spreading the roots on it, and covering them with more grass. In a few minutes they are thoroughly steamed, and are really excellent eating.

We named the river Ah-ne-pash—pretty name, isn't it? which it will probably retain, as it is embodied in our report.

We climbed a high peak above our camp, and had a pretty view of Ah-ne-pash valley, and the stream passing thro it—enlivened by sunshine, too, for—

"As I parted from the pine trees,
Closing in a gloomy crowd,
O'er a swell above their branches
From a whitening western cloud,
Sunlight broke upon the valley—
Filling with an instant glow
All its basin, from the streamlet
To the dark edge touched with snow."[69]

We travelled, next morning, down the valley, noticing great change in the timber. Very little cedar or fir to be seen, now—almost all yellow pine, the uplands and mountain sides covered with beautiful grasses. We arrived at Ahnepash River, and (crossing another swift stream) camped upon its bank. This, and [These are] the two first streams we see flowing into the Columbia, east of the Cascade Mountains. A few miles below was a fishery, where the Indians spear salmon that come up the Columbia River, above the Cascade Falls, enter the Yakima, (into which I think this new river falls,) and so on, until they come up *this* stream, and almost reach the base of the mountains, within fifteen miles of the headwaters of Greenwater, which empties into Puget Sound.[70]

Meeting Indians Along Ah-ne-pash River (June 13-16)

The trail separated at this point. The one we left where we crossed the last stream, went over the mountain to the left. That which we intended following led down

Edgar Rock and the Naches River. Photograph by Dennis Larsen.

the stream, crossing it twice. But in the morning, old whitehead rode up and informed us the stream was so high, we would have to swim our horses—so we reluctantly recrossed and took the other trail tho it afterwards turned out a fortunate choice. We continued our course along this valley, till a huge rock blocked up the narrowed extremity, and shut us out from the valley.[71] Here, again, our guides were at fault but Shazer, retracing his course, struck the trail some distance back, going over the hill—and so on we went, till night set in.

Here we noticed another quite large valley running west, and wondered where it led to—but our Indians could tell nothing about it.[72]—We crossed upon a high stony mountain, where a road would be impracticable; but, as the stream at its base is very low in summer and fall, we decided the road must run along the bottom land, tho it involved many fordings. Some distance further on we came where a ledge of rocks hemmed in the valley completely, and over this we had to climb and clamber in a manner more curious than pleasant.

Here "Old Bones," our pack-horse, slid off the ledge, in the act of turning around to wink at his following companion, (for whom he had contracted a friendship,) and went "pondhering"[73] down to the water, but checked himself by plowing his head in the hard gravelly road. I got down to him but dared not release the pack, as he lay in such a precipitous place that my strength alone would have been insufficient to prevent it from rolling into the stream; assistance having been rendered, he was released and repacked, sullenly rejoined the train, showing much mortification, and deigning no sign of recognition to his more fortunate companion.

This rock, the Indians say, is the *ne plus ultra* of snakes; hordes below, but not one above. I do not doubt it; it looks probable—cannot possibly see how a snake could wriggle himself up there.

Just at the foot of the rock was another salmon fishery and an Indian camp, with any number of its inhabitants engaged in pursuing the finny tribe.

They have many peculiar customs connected with their fishing, and have a high opinion of the salmon's cunning and wariness. Shazer came near "busting up the whole concern" by leading his horse to the river to drink. The Indians kicked up a terrible fuss. If the horse drank there, the water would tell the salmon, which would not then come up, they contended.

Soon after, Edgar's little child[74] took a tin-cup to the brink, but they wouldn't let her dip any water, and hurried her off. In my ignorance of the etiquette of salmon fishing, I was about to wash my feet in the stream, but succumbed to the prohibition, reflection suggesting such a process *might* possibly have a deleterious influence.

The Indians very energetically attempted to convince us that the salmon would "hum-m-m" to each other—so we had to get our drinking water and refresh our horses at a little spring back of us—and, as *that* flowed into the stream, we had to carry it some distance away, to prevent any running back into it. It saved us some labor, however, for they were so extremely anxious to have it done carefully, that they *did it themselves!* I went up to their fishery, and saw an old fellow, with a spear similar to those described as in use on Lewis River, pierce a salmon weighing more than forty pounds, which he killed by beating it with a club.

In the morning we missed our horses, and sent the Indians after them. During their absence I had a long talk with an old Indian of the village, named Peter, who had travelled extensively—been to California, mined there, accumulated a considerable amount of gold, but had met a "Spaniole" (as he termed a Spaniard,) who had walked off with his earnings, leaving Peter "for dead;" he had also roamed over all the country, between the Cascade and Rocky Mountains. I had told him of the extensive salt lake I had heard was somewhere there, and he described it as being *very* extensive, lying in the "Spookan" [Spokane] region, about ten days ride, (or 400 miles) from here, in a north-easterly direction. I presume he was correct in his calculations, for I got him to point out the direction of Vancouver, and Wallah-Wallah, which he did with accuracy, as my compass proved.

When the Indians returned with our horses, with their assistance we carried our "plunder" over the next ledge of rocks, and then repacking, travelled over little spurs of mountains, down into narrow river bottoms, fringed with clumps of elder, maple, and weeping elms—and over beautiful uplands, till even, when we came to a hill of long easy ascent, from whose high emnence we could see, spread out beneath our feet, the wide waving prairie, forming the first of a succession of them, alternated by uplands and rolling timbered lands, stretching 150 miles, to Wallah-Wallah—a most lovely view! nature, in its own wildness, beauty and loveliness—for, with the exception of our party, grouped upon the bluff and absorbed in its contemplation, there was no sign

"Of man or brute,
Nor dent of hoof, nor print of foot,
Lay in the wide, luxuriant soil
Nor sign of travel—none of toil
And not an insect's shrill, small hum,
No matin song, nor voice, was borne,
From herb or thicket."[75]

We lingered here awhile, charmed by the outlook, and the peace of the scene stealing into our hearts.[76]

An Indian overtook us, and invited us to put up at his ranche, where we could camp. Turning off into a little valley to the south, we found his home—a lodge of skins, constructed after the manner of the Sioux, and quite a large number of horses and cattle roaming about denoted thrift. Very hospitable we found him, bringing us milk and roots, and, in the morning, like a true host, he brought each of us fresh horses, and invited us out to see the country round about. We were all too much wearied to accept his offer. From him we learned that the valley we had observed far up the stream also led out into the prairie, and so avoided the most unpleasant part of the route down the river.

Allen and Shazer Return to Olympia (June 16-20)

Our journey was now at an end! our mission finished. We had ascertained, from personal observation, the practicability of cutting a road thro the Cascade Range, which had hitherto been doubted. It was universally known to be an open prairie and rolling country from here to Wallah-Wallah, so that it was unnecessary to go further, and very necessary to make our report as quick as possible, that the people of Washington might take measures to have it opened in time for the coming emigration of '53. It was therefore decided that two of us should go back "by express" the nearest way, and inform the people of our success, while the others should follow after more leisurely, and blaze out the road—also endeavoring to elicit more information from Indian guides they would procure to return with them on this side of the mountains. It was decided that Shazer and I should return and report, which suited me exactly, as I was getting homesick, and wanted to see the little cabin again, and, beside, wanted to answer William's letter, (which, by the kindness of the postmaster, had been forwarded by an Indian runner, previous to our leaving Edgar's) respecting the erection of a saw-mill on Puget Sound.[77]

Next day we were ready to start. After trying in vain to procure an Indian guide, and finding none who were willing to start for two days to come, we decided to do what no white man ever did before—return alone. Our provisions had become very scarce, and we were unsuccessful in our endeavor to procure some dried salmon from the Indians, who had none to spare, as the small-pox had prevailed to such an extent as to interfere with their fishing operations—and they were scattered about, in twos and threes, thro all the little vallies of the mountains, which afforded roots and berries for their subsistence.

They attribute the origin of the fell pestilence to the robbery of a store in Vancouver, supposed to have been perpetrated by "bad Indians"—and say that the proprietor invited all the neighboring Indians to a feast, and presented them many little presents, feasting them with sugar and every Indian luxury, at which they were much pleased, and returned to their mountain homes, when the dread disease broke out among them with great virulence.[78] I can hardly believe such a fiendish act of retaliation possible, but it seems not improbable, when one recalls Sublette's devilish revenge (on the Sioux who robbed *him*,) by causing a mule load of inoculated blankets to be exposed to their depredations.

Our stock of provisions had necessarily to be very light, and consisted of ten pounds of flour, four baked cakes, and about two pounds of pickled pork—and this was to last us for a distance which had required us two weeks to travel! We determined to eat sparingly, and ride hard, trusting to chance to procure more, wishing our comrades (who would have robbed themselves to spare us more,) good fortune, and they bidding us "God speed" as we dashed off.

After travelling some miles, we overtook a small party of Indians going to the west of the mountains, who would have been of service, but their travelling was too slow for our motions, and they were soon in the rear.

Travelling rapidly, we came to Ahnepash River, where we saw a tremendous black bear crossing the stream before us. As Shazer had my Colt's Revolver, he ran down to shoot it, while I secured our horses; he got within eight yards of him, before he was discovered, and would have shot him, but the pistol snapped; he dashed thro the underbrush, and Shazer kept popping at him in vain. Seeing he was coming up the part of the bluff I was on, and likely to frighten our horses, I mounted mine and dashed down to "head him off," but my horse had smelt him and was in no particular hurry to meet so rough a customer—so, in answer to my spurring only curvetted,[79] so as to threaten to throw me over the bluff into the river.

I was too late; he had taken the bluff side, and was going up "like a streak," over those sharp-pointed volcanic rocks, yet as gingerly as if treading on Brussells carpet! What made it more aggravating was that Shazer had, till that day, carried a gun that had belonged to Edgar, but left it behind that morning, as being too troublesome to carry.

This little episode over, we resumed our march, thro the little prarie where, (I had forgotten to mention,) in coming over before, our advanced guard of Indians had scared up a herd of forty or fifty elk.

I need not relate to you all the "incidents by flood and field" that befel us on our return—Suffice to say that, with our compass, we struck a straight course, and crossed the snow as true a "bee line," finding the tops of the saplings we had cut, on our progress over the mountains, level with the snow, were now five feet above it, showing the sun had melted it that much in the interim. We also found a horse of Edgar's, which we had abandoned in going over, lying dead in the snow. It was raining, as usual, and I concluded it had a contract so to do, as long as the snow lasted.

We camped on Bear River, and had a glorious night's sleep, wet as we were, despite a dismal rain all night long. Made a hearty breakfast on boiled flour and water, with a "dub" of pork in it for flavoring; too hurried to cook it, we ate some raw. The remainder of our way was comparatively easier, and on the evening of the fourth day we slept at Sandy Smith's. My horse (the wild one) I turned loose without check or lariat, on the same prairie on which he had given me so much trouble before; in the morning found him not far distant, and he came at my call, to be bridled and saddled, like a good old animal that had "had the starch taken out of him," as he was, much to Sandy's surprise.

Sandy Smith was delighted at our favorable report on the practicability of the road. Called at Edgar's, told them of our success; horse hardly able to be got out of a walk; rode seven miles out of my way, to Nisqually, where Shazer promised I should have a fresh horse; wanted to go into Olympia in style. Gave him, and his neighbor Packwood,[80] some of the seeds you sent me; he has a fine farm. There is a singular prairie here, which goes right to the bay, and in June tides (the highest in the year,) it is partly covered with water. He has 1,000 acres of good rich land. He married a girl of fourteen, coming over the plains, and has an uncountable number of children all apparently of the same size and age.[81]

Here, thro Shazer's kindness, I procured an American mare, which he avers to be the fastest in the teritory. He having also a fresh horse, driving our tired horses before us, we rode like madmen to Olympia, only eleven miles, putting them thro in a style never before seen in that country, for

"We cared not for stumps, and we stayed not for stone;

And dashed over creeks, where bridge there was none."[82]

At the outskirts of Olympia, (and very extensive those "outskirts" are!) we met a party of ladies and gentlemen, going out to Judge Y's;[83] dashed thro them, scarce reining up to tell them the road *could* be made—hurried on, the sound of their cheer following on our heels—pitched into Olympia, almost to the other end of which we had to go, ere we could check our excited horses. *Such* a commotion as our arrival kicked up, in that extensive place! We immediately delivered a brief report to the committee.

Road Viewers' Report

The *Columbian* printed the report in its July 2, 1853 issue, which until now has been the only account of the road viewers' journey available to historians.

Report of the Cascade Road Exploring Party. *Near Yakima Valley*, June 18, 1853. *To Messrs. Moore, Hurd and others*—In accordance with the instructions given through you, as a committee of arrangements, by the citizens of Washington Territory, to us as a committee to view out a practicable wagon road from Puget Sound across the Cascade mountains—having crossed from the Yelm prairie to Yakima valley, by a trail some 25 miles north of Mount Rainer—we now lay before you the result of our observations.

We supposed when leaving Olympia that the most feasible route would be that described by Mr. Blanchet in his journal of 1848, but upon reaching John Edgar's we were assured by the Indians that the trail was then impassable on account of snow, and that the trail spoken of above was the only feasible one.

June 5th, we left Yelm prairie, and, crossing Nisqually river, and traveling principally through prairie, struck Puallep river some three miles above where the old road crosses. Here the road will leave the trail, and striking up a small stream to the right, cross White river some four miles above the crossing of the old road—follows up the valley of White river until it reaches Greenwater, and leaving White water, leads up the valley of Greenwater until it strikes the trail again at the base of the mountain. Here it will lead up a kanyon to the left of the trail, and rising the mountain and finding no very difficult hills, descends by a gentle slope to Ahnepash river, the first stream flowing into the Columbia east of the Cascade range. Follows the valley of this stream to where it empties into Bumping river, and leaving it strikes up a valley to the left leading up on to the dividing ridge between Drift and Yakima rivers. From which ridge to Walla-walla the country is open prairie.

Knowing that it is necessary to make our report as soon as possible, we have resolved at this point to separate, and two of us return immediately to Olympia, leaving the others to follow after, blazing out the road; and having procured good Indian guides, probably make some material improvements upon the route mentioned. At the time of their return we will make out a report in detail for your further satisfaction. We would only say at present that we deem the road so feasible that we consider it unnecessary to survey any other trail, knowing that the time for the completion of the road in time for the immigration of '53 is already brief enough. The whole distance will probably be about 80 miles from Muck river to the dividing ridge above mentioned.

With our limited time for observation, this route is the only one we can follow so as to avoid the high hill of Puallep, and so we were unable to bring into line any portion of the road already cut. The whole of the proposed route is well watered and affords very good grass. The valley of White river contains many most beautiful and rich prairies—and the summit of the mountain being almost a continual chain of prairies.

We lay the above condensed report before you, until such time as the return survey will enable us to enter more fully into detail.
John Edgar,
Whitfield Kirtley,
Edward J. Allen,
George Shazer.

All that remained was to actually build the wagon road.

Logging in a Northwest forest, as Allen would have done on his land claim on Budd Inlet at Olympia. *The West Shore*, December 1885.

Robbed!

Allen returned to Olympia from his adventure as a road viewer and delivered his report to the road committee. He next made a trip to the post office to collect his usual pile of mail, and then returned to his cabin on Budd Inlet only to find it had been burglarized during his absence. His July 20, 1853, letter continues his narrative just as he had finished delivering his report.

A Dastardly Deed

I lingered around, modestly, to answer all questions, and then rode to the Post Office; a barrelful of letters, of course. Went down with these to Brooks,' as my canoe was not up, and a neighbor, going down, promised to land me there. Remained all night at Brooks,' reading the letters, about half of which I got thro, and next morning (with Dr. Eggers) paddled over and ascended the bank. All things looked natural, *exteriorly*—the eagle sat in his usual place,

...."on a swinging limb;
He winked at me, and I nodded at him."[1]

—as I ascended. The sunshine glinted pleasantly thro a vista, our labors had opened in the dark woods. Our old cat relaxed herself from her upright position, watching to tom-tit on a neighboring twig, and almost tied herself in a knot in his exuberance of welcomes! So far, all right—but, entering the house, I saw at once it had been *robbed!* ya-as—ROBBED, despoiled, pillaged, plundered!

Some villains, pirates, fillibusters, in my absence had made sad havoc, and scattered my "household gods" about, in a worse manner than ever were Byron's. *All* the provisions laid in for a month—*that* I didn't care for; all my shirts and stockings—that I didn't care for either, as they could easily be replaced—not so easily, either, for they were "home-made;" the little store of needles, &c, which will put me to some inconvenience; the razor they're welcome to, I shan't miss it, but I *would* like to see them "try it on!" All these they took—with a knife William gave me; and what makes me think it was Indians "done the deed" was my missing a copy of the *Spiritual Messenger*, which an Indian would naturally appreciate, as, from the striking similarity between its sentences and the Chinook jargon, he would suppose it was something "in his line."

{Of course I want to blame it on the indians, for sentiment sake, but if I really went on the hunt for them, I would paddle down the Sound, and enquire what nomadic white men had passed down the Sound; it was really the more aggravating as we do not expect such actions, and are in the habit of trusting much to the general good faith. We do not among our tribulations include thieving.}[2]

Worse than all, and all that I regret, is the loss of the Christmas ring,[3] with the motto, dearer than ten times its value; and the gold sleeve buttons, (worn ten years,) that William sent me.[4] No money can replace them, and the only comfort left me in my affliction is the hope that, from their being so easily recognized, and the characteristic habit of the Indians, of using the nose as a feature to hang trinkets on, I may some day recover them.[5]

This diabolical outrage was committed within three days past. Dr. Eggers (a new personage on our little stage, of whom more hereafter,) was over here then, and found all right. No one knows aught about it—altho I think "Tom," the cat, could enlighten us if he had chance to "say his say," as he seemed particularly agitated, excited, and indignant, during the progress of our investigations into the extent of our loss; and has been observed since, on one or two occasions, when Indians come with salmon or oysters, to be quite frigid, if not insulting, in his demeanor toward them. I am convinced that no "white man" in Washington Territory would be guilty of it—but must bide my time, with the sympathy of the whole community, in the hope that I may yet recover the ring and buttons, at least.

Three nights since my return, and how beautifully breakt the morning of the third day!

"Floods of more golden sunshine
Never the eye saw rolled
From temple and porch, or cold gray spire
That turns in the light to gold!"[6]

I have not been to town since my return, tho the citizens are to have a meeting to-morrow, and discuss the practicability of the road being cut by voluntary contributions, if government will not assist—and, as to-morrow is the anniversary of our national independence, and much interest is felt in the proposed road, there will no doubt be a large attendance.

In the three days spent at home, after a month's absence, I have completely explored it. Changes have occurred meanwhile. The men employed are all away, having under Ensign's superintendence, completed the 5,000 feet of lumber contract, a week after I left. Ensign is on a visit, down the Sound; another of our men is on the next claim below; there are only three of us here now, and we are provisioned afresh for some time.

Dr. Eggers

Having alluded to Dr. Eggers as a new character in our drama, you may wonder how he came here. He wonders, himself. Perhaps, among his other peculiarities being somewhat of an epicure, the fame of my cookery may have reached his last residence, and a wish to compare experience brought him over. Be that as it may, he came, and is now with us; how long his erratic and restless disposition forbids even a guess.

{The Dr has the German temperament, and philosophically sees and makes no differance in any honorable occupation. I see this, have been taught it here, where it has been brought home to me, but the Dr has always felt so. He has had a very liberal education; is a good chemist, botanist, geologist, and conversant with the sciences generally, and in a thorough way; and tomorrow did it please him would cook for a lumber camp, give music lessons, teach chemistry, and out of ordinary things, make the necessary appliances for the laboratory, and do all well and with equal acceptance, efficiency and industry, and withall is unconscious of his large ability, and is one of the most modest of men through his utter unconsciousness. He has done all these things, and will probably do them all again, and might at last settle down into a college professor with all his wealth of experience and knowledge of all kinds of useful occupations and knowledge of men, which with his German student habit, he will not make personally useful for his own advancement...He has done every thing, from practising medicine to cutting tobacco and making earthenware, and if he practiced medicine, depend upon it he was a medical graduate, and if he made earthen ware, he brought to it some chemical knowledge that made him valuable... Now you see a man like that is a delight to have in the presence.}[7]

I had, on my claim, a patch, where we have got out timber, nearly free from it—and proposed to him to clear it entirely, and I would furnish the potatoes for seed, and give him two-thirds of the crop; so he has been steadily at work and has now from half to three-quarters of an acre ready for planting.

Many, who read these letters, may know him as he was from Pittsburg, where he taught music for some time, being an excellent musician—a scholar, familiar with geology, botany, chemistry, and most of the sciences—thoroughly German in his temperament, the most independent man I never knew, and one who has seen more changes and chances of life; having undergone almost as many transmigrations as "Indor"[8]—been all over the Western States, and (since his arrival) traversed a great portion of Oregon.—Every night he has made me laugh, in listening to his queer adventures, told in quaint, broken, comical way. He has done a little of everything, from practicing medicine to cutting tobacco and making earthenware. A very great accession to our little circle, is the doctor, I assure you!

Thoughts on the Neighborhood

It is confidently expected, by the people generally that upon Gen. Stevens' arrival Olympia will be made the seat of government of Washington Territory. I know of no more suitable place, being as you may say "at the head of navigation," and commanding all the country around, on both sides of the Sound, and below to the Columbia River. The communication with the interior will have to be thro the straits of Fuca, unless a route for a railroad is found, leading from Gray's Harbour up the Chickalees [Chehalis] Valley, to the upper part of the Sound, near Nisqually. This of course, would give a fresh impetus to all places on that portion of it, and particularly Olympia. All claims on the Sound will be valuable. I am more and more pleased with my selection; and, seated at the window, inditing this letter, with the glorious outlook before me, no longer think my neighbor's valuation of his claim,[9] above me, at $5,000 (which he has been offered,) extravagant. No money can buy mine—that is, with the surroundings included. During my absence an order was issued by the court, to open a road "from Yellow Prairie to the Sound, thro Allen's claim," which will materially add to its value.[10]

How still and quiet seem everything around me, as I write! The bay which yesterday under a high wind, was tossed into waves, tossing up their sharp white helmets, to-day is calm and unruffled as an infant's slumber. Far up, Olympia seems nestled upon the water, as quiet as a pictured city. Old Rainer looms up, stern and grand, sixty miles away, with a noble presence. Two weeks ago I was under his shadow, not more than fifteen or twenty miles off—and he seemed equally distant. The slight breeze, bearing the breath of a thousand flowers, after crossing his snow-crowned peak, cools my brow.

Across the bay Mr. Wylie is "clearing," the wreaths of blue smoke languidly coiling upward from his labors. The grouse are calling again, after their few days silence, their deep sonorous notes swelling on the wind. The mail canoe[11] to Nisqually is passing down, the Indian oarsmen singing, as they row, a strange wild melody; following directly after comes the sloop Sarah Stern, her sails in the bright morning sun gleaming white, seemingly taking across the bay as if coming directly to our inlet, perhaps for our cargo of timber.

{The Indians chanting a strange weird melody... as their ancestors sang, so sing they to day, the paddles keeping the same time on the gunwale of the canoe as awakened the echoes ages ago.

{In the sameness of things here, one always seems to stand amid the dead centuries. With the white races, the past has but little place. With the red man, it is abiding... I do not think the indian ever comes near seeing the light that never was on sea or land but he must have some conception of what underlies, because he invests natural things with spiritual attributes, and considers them Gods, and also has his visions in puberty that influence his life, and has his poetic traditions of events.

{I am in grievous doubt, whether he has much else to offer as classing him above the brutes, and I am sorry to say so. The indian of Cooper[12] was a beautiful fiction. I think no one could ever been justified in a full faith in an indian. The centuries have too strong a grip on him.

{Sooth to say, if he started with any reverance for the white race, as unquestionably some of the eastern tribes did, it was lost because they became aware that the profession of the white man did not accord with his practice as much as did the red mans. He had not reason enough to consider, that such slight theology as he had, was based upon a large acceptance of human nature and was adapted to it, and required but little restraint to keep within its limits, while we set up a standard so high, that Heaven offered no higher, and human nature could not reach it.

{The indian could scalp and torture, and be within the limits of his religious belief, which indeed seemed to demand such treatment of an enemy, and he was extremely conscientious in fulfilling his utmost duty in such cases.

{When the white man was met by any temptation to inflict punishment or avenge himself, he had to restrain his natural impulses and conduct himself on a divine basis, and of course he fell short.

{The indian did'nt fall short, and didnt see any reason why a white man should, and lost his respect for him.

{Religously the indian had infinitely the easier job, but he did not recognise that.}[13]

On such a day as this, the like of which you never see in "Auld Reekie," I would like you to see Oregon—and especially my individual share of it on Pugets Sound—verily would you be tempted to say: "Behold mine eyes have seen thy glory, let now thy servant depart in peace. Selah!"[14]

One small remaining segment of the Naches Wagon Road may be found in Federation Forest State Park, three miles west of Greenwater, Washington. Photograph by Dennis Larsen.

Building the People's Road

The story of the building of the Naches Pass wagon road and of the first wagon train to use it has achieved nearly legendary status in Northwest history.[1]

The emigrants of that wagon train complained bitterly of finding only a slight track through the Cascade Mountains of Washington Territory and of having to hack their way through the forest. One member of that wagon train, George Himes, who was then nine years old, told of coming to a cliff down which the wagons had to be lowered by rope. According to Himes, the pioneers were short on rope and were thus forced to butcher and skin three oxen in order to make their hides into additional rope. This story has been told time and again in Northwest history, but it may be just that—a story.

Ezra Meeker repeated it in his 1905 book *Pioneer Reminiscences of Puget Sound*.[2] He devoted several chapters to the building of the road and to the travails of that first wagon train. Over the years Meeker and other historians also claimed that the road builders gave up their project prematurely and returned to Olympia after hearing from an Indian that no emigrants were coming west over the trail that season. Thus the wagon train was left to find its own way through the mountains. This erroneous claim was based on a statement made years after the event by James Longmire, a leader of the wagon train.[3]

Until now very little has been heard from the man in charge of actually building that infamous road. A few accounts in the *Columbian* (at the time the only newspaper on Puget Sound) and a brief discussion in the appendix of the 1913 version of *The Canoe and the Saddle*[4] summed up Allen's comments on the project. Allen's 1853 letters and the Pittsburgh Manuscript provide two new lengthy versions of the story giving great detail and a different perspective on the events of that summer.

Organizing

Allen's account of building the Naches Pass wagon road began in a letter dated July 20, 1853, which described the delivery of the report on his scouting trip over Naches Pass.

On "the Fourth" we had a great day in Olympia. I went up in the morning, to attend the meeting called to receive our report and devise and discuss measures to have the Cascade Road opened. There were more people gathered than I had ever seen assembled before—all interested in the measure. The report I had handed in was necessarily a brief one, as the other party[5] was yet to come in, but it convinced all of its practicability—and it was the resolve of the people and the decision of the meeting that it should be done, at their own expense if government refused to aid. Capt. M'Clelland[6] was daily expected with a force, to meet Gov. Stevens coming over [the Plains], and before adjourning the meeting appointed a committee to wait upon him on his arrival, and ascertain if he would not co-operate, and approve the route selected by the viewers.[7]

The meeting over, the majority (if not all) adjourned to the Methodist Church, the largest room in the place, accompanied by a large number of ladies, to celebrate the Fourth, which took up nearly all day, and was done up in a manner equal to anything of the kind in "the States"—and more than that there was less humbug and cant, shorter speeches and more sensible, fewer allusions to worn-out themes generally discussed at such times, wherein "our glorious Union" is represented in great danger, from the machinations of the party the speaker may be opposed to. The toasts were generally local, the prosecution of the Cascade Road being the most popular, while the presence of so many of the gentler sex, gave a social and refined tones to the whole affair.[8]

Another meeting at Olympia on the 9th. I called to hear the report of the Committee appointed to wait upon and confer with Captain M'Clelland respecting his approval of and assistance in cutting through the Cascade, or "People's road."

As Captain M'Clelland had not yet arrived, the committee had no report to make, but to urge (which they did) the appointment of another committee, (who were elected unanimously,) to proceed immediately to collect subscriptions and enroll volunteers to cut the road themselves, and trust to the generosity and liberality of the Government to reimburse them afterward, out of the $20,000 appropriation never used, the committee to use every endeavor to prosecute the enterprise, so as to be in time to bring over the emigration of '53. I was placed on the committee,[9] and foresee a week or two's riding through the country involved, and a further absence from my "little housie."

Collecting Subscriptions

July 17.—Have written nothing in this journal kind of letter, since the 10th, having during that time been very little at home, but scouring the country, in performance of the duties imposed on me by the meeting of July 9th, as one of a committee to collect subscriptions to prosecute the new road. We have been remarkably successful, and there seems to be but one universal feeling—to give liberally to that object; and you must give them credit for other and better motives for so doing than the mere advantages likely to accrue from its completion, numerous and great as they will be, diverting a large emigration from Southern Oregon to this territory. Almost all who have selected Puget Sound as their residence, have a vivid recollection of the hardships and exposures "coming over the plains"—and the additional suffering to be endured, and obstacles to overcome, in their further progress from the Columbia River, over the pack trail. Many of their neighbors, friends, or relatives, are annually coming over, and their philanthropic and praiseworthy desire is to alleviate their suffering and lighten their sorrows by shortening the way. Ere this the plains have been dotted by the white-covered wagons of the coming emigration, and round many a camp-fire are collected groups of emigrants, talking in half-suppressed tones of the homes they have left which seem already so far distant—building up visions, too, of future happiness in our goodly land. As yet, they know nothing, comparatively, of the hardships—travelling up the broad valley of the Platte, they dream not yet of the long dreary desert, rocky hills, and dusty plains which lies beyond. Upward and onward, they press—until the gates of the Sweetwater shut out all the bright valley behind, and open a dreamy vista beyond, of toil and suffering they must endure, ere they arrive at their destination. Many will faint by the wayside, sleeping the sleep that knows no waking—and, with brief delay, small time allowed for tears, the caravan will move on, the rearmost companies stopping carelessly to read upon the rude headboards the names of the sleepers; until at last, the poor rear will have passed, and in the oppressive silence and solitude of the wide prairie their rest will be unbroken.

We had another meeting, yesterday [July 16], at Olympia, to report the cheering news of our collections, and enrolment of volunteers—some giving their personal services, and others (whose business prevented them from going,) furnished outfits for those who would. From all appearance, there was sufficient force to open the road, and enough provision, not only to sustain those engaged in it, but to furnish a surplus to forward for the coming emigration. The meeting, urging further collections, and resolving to await for one week longer the expected arrival of Capt. M'Clelland, so as to secure his co-operation, adjourned—determined, the next week, to appoint leaders for the expedition, and push it thro themselves, as a public measure, which could no longer be delayed, and in which all were interested.

A more entertaining account of obtaining subscriptions, written years later by Captain John G. Parker, follows:[10]

During the summer of 1853 the residents of Olympia received word that there was a large party of emigrants with ox teams, prairie schooners and horses near the confluence of the Snake and Columbia rivers, who, with great difficulty, were endeavoring to come to the Sound and willing to try the mountain trails. The citizens of Olympia and vicinity were desirous of extending a hearty welcome, and E.J. Allen called on me with a subscription list to raise funds necessary to cut a wagon road across the mountains by improving the Hudson Bay company's trail via the Natchez pass. The subscription then comprised the signatures of G.A. Barnes, Joseph Cushman and Moses Bettman, pledging $25 each. I persuaded Mr. Allen (now Major Allen of Pittsburg, Pa.) to be guided by my suggestions, to which he assented. Destroying his subscription list, I wrote a new one, headed it with $100 and sent him with it to Moses Bettman. Moses' pride was instantly aroused and down went his name for $200. George Barnes raised Bettman's $200 by subscribing $300, and Judge Cushman, not to be outdone by Barnes, subscribed $400. Captain Crosby also subscribed liberally, the amount I have forgotten, and Dr. W.F. Tolmie, then chief factor of the Hudson Bay company, also, which was greatly appreciated by all of us. Other residents in the vicinity donated provisions of various kinds and other necessary material, and a party of over fifty muscular young men from Olympia and vicinity fitted out with the supplies and equipments needed, commenced at Montgomery's on the eastern edge of the Nisqually prairie and cut a rough road across the mountains. Quincy A. Brooks and myself met the emigrants on the summit of the Cascades, near the Natchez pass. We were on horseback and attended by the celebrated Indian chief and guide, Quiemulth [Quiemuth], Leschi's brother, leading a pack horse that carried our blankets and provisions. Among the party of emigrants that we met were: Charles Biles and wife, their sons David [and] Charlie, Van Ogle, Asher Sargeant and family, Nelson Sargeant, Marion and sisters, Matt Baker and family, Isaac and Abram Woolery and their families and Himes and family. They were followed a couple of weeks later by another party, among whom were Isaac Carson, Rev. G.F. Whitworth and family, Mr. Boatman and family and the Wright brothers.

A few discrepancies from the known record appeared in Parker's account. The subscription amounts he listed were different from those published in "Receipts."[11] Meeker identified thirty-four members in the two road building parties. However, a number of Indians and others guiding and taking care of the stock could easily have swelled the number to fifty. Parker was also mistaken about the Boatman family: they came to Puget Sound via the Cowlitz Trail, not Naches Pass.[12] Finally, no contemporary account of Brooks and Parker meeting the emigrants at Naches Pass has been found. However, the *Yakima Herald* of August 15, 1889, in a story titled "An Old Pathfinder," stated that Parker carried tobacco and women's shoes to the summit and "made happy the hearts of the men who had gone hungry for the weed, and gladdened the hearts of the women, whose shoes had worn out on the long march." Meeker also gave credence to Parker's story. "Accompanying the party of road workers was Quiemuth, a half-brother of Leschi, who acted as a guide and led the horse upon which were packed the blankets and provisions of Parker and Allen."[13]

Edward Jay Allen's July 20, 1853, letter continues:

July 25.—Home again, after four or five days absence, collecting more men and money. Have ridden over a good extent of country, till my little note-book is plethoric with jotting down what I have seen at Fort Stilacoon [Steilacoom], Nisqually, and other places—descriptions of all which will be inflicted upon you on the first opportunity, doubtless to your great dismay.

There is going to be another break up on our little peninsula—which will, for some two months or more be left to the sole occupancy of the eagle, crows, and occasional aboriginal visitors—for, having waited a week in vain for Capt. M'Clelland we can afford to lose no more time if we wish to benefit this year's emigration. My former compatriot, (Kirtley,) leads one

This 1857 sketch depicts the city of Steilacoom where the road builders outfitted for their journey into the mountains. Courtesy Tacoma Public Library, C137336-8.

party, and has already gone to the Yakima, to work westwardly. I start, bright and early in the morning, from Olympia, to work eastwardly, with eighteen or twenty stout hearts and willing hands—all good men and true, and am only down now "to pack up" (as I don't want to be robbed again!) what effects I have, and leave them with my neighbor [Adam] Wylie,—together with my faithful cat who has in vain besought to be allowed to accompany me.

There will not be a soul left on the place, as one of the men goes along and I have sent [Shirley] Ensign on ahead to the Columbia, to see if he cannot build some kind of a boat for a ferry there. I am not afraid that my claim (requiring a two years residence on it to perfect a title,) will be "jumped" in my absence, as the citizens of Olympia would protect my interests while I am attending to theirs—and besides, we have no men here mean enough (as in California,) for such a trick; they are of a different stripe, altogether—and have added another commandment to "the ten" a[s] equally obligatory and attended by equally a severe punishment: "Thou shalt not "jump" thy neighbor's claim!"

I expect to be absent about two months, and will be so incessantly and arduously employed during that time, that I shall not have time to write, even were there any "mail facilities" in the region I go to—so it may be a long time ere you hear from me, and "by the same token", it will consequently be as long an interval ere I hear from you! and this thought rather saddens me, as I sit writing—the shadows thickening around...and unwilling to break the spell, I linger as I write the word......*Farewell!*

Fervently yours,
Edward J. Allen

Hard at work building the wagon road, Allen was not able to write his family until November.

"Allen's Claim," near Olympia, Puget's Sound, Wash. T., Nov. 3, 1853.
Mr. Wm. H Allen, Steamer Norma, N.O.:[14]

Dear Brother—It seems an eternity to me, and doubtless longer to you, since I wrote you a good old-fashioned letter; and I think of it with remorse, as I reflect that amid all your engrossing duties, the journal-like monthly letter of from five to ten sheets has never failed reaching me. I have written home oftener, but briefly, and kept up my correspondence with my Pittsburgh friends, knowing that the "home folks" would, through them and from our papers, be informed of my wherabout and projects.

The truth is, I have been so arduously and incessantly occupied, and for months been so far removed from all mail facilities, that it has been impossible for me to write to you sooner. When I wrote you last, (I think it was July 27, with 18 men—Mr. Kirtly my compatriot in the previous survey of the road, having already preceded me with his party, to cross over to the Yakima valley, and work toward me, a $1000 voluntarily subscribed, a wagon load of provisions, and some mules I had procured from the Fort, (demanding them "in the name of the Great Jehovah" and the people of Washington Territory,) I started to cut the "Cascade Road."[15]

Our expedition started from Nisqually, which is situated about 25 miles lower down the Bay.[16] Back of it are what are called the Nisqually plains, the locality of the scenery Wilkes[17] describes so graphically and enthusiastically. These plains are prairies extending 20 miles up and down the bay, coming within from one to three miles of the water, from which it is separated by a belt of beautiful, park-like woods and groves.

These plains are generally very gravelly, and yet are covered with a fine luxuriant growth of grass; at the time we passed over them, from the summer being dry, yellow, but affording rich pasturage to the cattle of the Hudson's Bay Company, some 10,000 of which have been for years roaming on these plains feeding only at night, or early in the morning and late in the evening, remaining in the woods all day.

The soil (all of it) is not generally considered tillable, except in the small bottoms margining the small streams which run through it; so the wild hordes of cattle roam through it, untended and unsheltered. It is fast being taken up for stock farms, and I hear of no more desirable locations or more rapturous glimpses of scenery. Wilkes in nowise exaggerated in his description of the country round about. It seems almost the work of art. The grouping of the "mots" of noble trees, dotting the plain here and there, the "upland brown," and silvery streams "fringed to their willowed margin."

About a mile and a half back of Stillacoom stands the American Fort; ten miles further up towards Olympia stands Nisqually, the Hudson Bay Company's fort, like Stillacoom, not placed on the bay, but on the plain about three miles back.

The Hudson Bay Company claim about 12 miles square, which comprises not only the two town sites of Stillacoom, upper and lower, but also the ground on which our fort is built, and for which we are silly enough to pay an annual rent of about $7,000.[18]

The settlers and emigrants have taken up claims and settled all round, paying no regard to their assumed rights. I have never paid much attention to the subject, but think myself their claim is groundless. I have never seen the treaty, but am informed it only guarantees the title to the lands that they really occupied, to the "Puget Sound Agricultural Company,"[19] which was inconsiderable in comparison to the amount they have since "annexed" under the pretence of their cattle having strayed over it, which seems a very frivolous ground, and it would be difficult to ascertain what extent of country they did range over, as there is no limit to restrain them, except the Nisqually river on the one side and Puyallop on the other, either of which are easily forded.

After crossing the plains, we came to where our wagon was stopped, and could proceed no further, and were therefore compelled to shift the burden to the backs of our mules—not without an exhibition on their part, of various circus performances, endeavours to make statues of themselves, by rearing upon their hind legs, (of which they have two,) and remaining in that position till fatigue would compel them to change it—evidently laboring under the impression that they could {decide themselves whether or no, they should carry their burdens},[20] which idea after some time and trouble we convinced them was absurd—and, all things being then satisfactorily adjusted, we deployed into line, and began to ascend the hill, presenting a picturesque and mayhap a grotesque appearance.

I have no time to give you an account of each days' progress, and the manner of our working.

"We forded the river and clumb the high hill,
Never our steeds for a day stood still—
Whether we lay in the cave or street
Our sleep felt soft on the hardest bed;

Or stretched on the sand, on our saddles spread
As a pillow beneath our resting head—
Fresh we arose on the morrow.
All our thoughts and words lent scope;
We had health, and we had hope,
Toil and trouble, but no sorrow."[21]

The last word must be excepted to, however—if anxiety and harassing cares are its principal components; for, so soon as operations commenced, the troubles began, the variety and extent of which it would be an impossible as well as a disagreeable task to describe. Every day brought forth fresh cares, and every evening saw us safely thro them. It was a free wild life—primitive and natural; we were out among the "eternities and realities."

Day by day we toiled on thro the mountains—treading where man never trod before—thro deep glens of heavy underbrush, which we cut away as we proceeded—our camp fire at night gleaming a home, welcome to me, and the advance guard; day by day still onward—opening the dark old woods, the growth of centuries—the narrow line of light marking our progress; now rising over small spars of mountains, defiling thro passes in deep caverns—still onward. Here, each well known locality of our previous viewing loomed up to us as goals to be reached.

Giant Trees

Allen's Pittsburgh Manuscript provided a much more detailed look at the life of the road builders in the field.

We had quite a caravan some nineteen in all, with a string of pack mules carrying our provision, blankets, cooking utensils and general "plunder." It was an altruistic venture, for our provision was contributed, the pack animals were lent by the U.S. thro Major Larned of Fort Steilacoom. The labor was voluntary, and there was enough cash in the hands of the committee at Olympia, to ensure all necessary future supplies. Dr. Tolmie for the Hudson's Bay Co had not only made a liberal cash subscription, but had agreed to permit us to take what cattle we needed from his herd, we to account thereafter at the price of $5 per head.

Andy Burge was the Packer, and brought us out many carcasses of tender beef, and never missed connection, being an indefatigable fellow.

As this movement was an epoch in the history of the new territory I will append later a list of the workmen under my charge, and also the contributors to the fund.

It was a little problem to me when we started out how long these men could be controlled under what would seem a slight cohesion, but though some of them naturally grew impatient when the task grew greater than their first conception of it, they were all steadfast and remained until the end.

I think none of them ever worked longer hours, or did more work in such hours than these men did. There are a great many hours of daylight, in twenty-four hours of a Washington summer day, and we worked the greater number of them. The general route was determined by the topography but there was an immense amount of detailed work in deciding the best road; with my little hand axe I was generally out an hour before the men, triangulating the whole valley (without instruments) and determining each rod of the way and having determined, marking all trees that would have to be felled. A direct route, even when otherwise possible, had to be governed by the timber. The trees would average six feet in diameter and many were ten or twelve feet or more. To fell such monsters was a costly and laborious feat and was to be avoided where possible.

Such trees when down measured nearly three hundred feet in length and to clear them out of the road was almost a great a task as felling them. And it was a question of judgment how to avoid them. If the test of measurement by rings of growth is received, some of these monarchs were on the earth when the Savior was living. It seemed an inexcusable thing to cut them down. The difficulty of doing it was a greater protection to them than the outraged sentiment.[22]

Robert Moore's War with the Hornets

One of our greatest troubles was the innumerable wasps nests, and hornets and the bees underground, there were so many hornets nests pendant from the lower limbs of the trees and shrubbery along the streams that it was often necessary to detail a man to destroy them a day in advance of the workmen. A twenty foot pole with a bunch of inflammable material on the end settled the matter, and in a morning a man would destroy hundreds of them of all sizes. It had to be done at least twenty four hours in advance and even then the woods were full of wandering insects only too ready to take offense, while searching for their lost domiciles.

It was best done in the late evening when a majority of the occupants were at home and perished with their habitation.

Robert More was detailed for this special service and developed a speculative desire to determine how close he could approach the nests while burning them, and continued to shorten his pole until it came almost to a hand operation. Continued warning did not affect his ardent desire to decide the matter, while the indignation of the hornets was growing stronger than their surprise, and it culminated in a vicious attack from the special nest he was destroying, in which all the homeless hornets which had previously been objects of his attention seemed to take part. His long hair was filled with them, all bare places were covered with their legions, and from far and near, they came through the air like bullets, striking the man with an absolute impact, who became frantic, and tore through the woods regardless of direction, knocking against pendant nests, and calling out new enemies, every minute. He was blind and deaf, but not at all dumb, he had frantically torn off all his clothing as he ran, in a desperate endeavor to rid himself of the insects which had penetrated every inch of his clothing, and entirely naked he had despairingly halted on the creek bank, standing over a bees nest in the ground, with a pendant hornets nest overhead. Unquestionably he would have been killed there, but gathering together heavy bunches of brush, several of us, not without many bitter stings, rushed in and pushed him over the bank into a deep pool and then falling flat on the ground lay quiet until the excitement had somewhat subsided and submitting to a few stings from such of the enemy who did not fully accept the usages of warfare, crawled slowly away. A hornet does not attack by lighting gingerly, and then hunting for a point to penetrate, but he goes through the air like a minie ball,[23] and strikes heavily and commences work with his business end before ascertaining whether the surface is penetrable, if he meets with discouragement he then tries another place and is disposed to fly off a little distance and come back with momentum. He seems to have a blind berserker rage.

In my little experiences in the advance, I had many stings, and had learned that in the hornets mad attacks, either he did not hazard risk of obstacles in flying low, or else his enemies had been higher up, and he seldom flew within two feet of the ground, so instructing my men, we prostrated and saved ourselves, with the exception of attack from a few radical fellows who did not respect the usages of the race.

There are always just a few such fellows to destroy the value of an accepted formula. It isn't confined to hornets.

While I lay there submitting to the perforations of the guerillas, I recalled the story told me by an Indian who was one of our guides in the exploration of the pass. At one time his blanket slipping, exposed a back so dreadfully excoriated that I was led to enquire what had caused it. The man had some years ago, suddenly encountered a grizzly bear with cubs, and attacked before he could escape, had thrown himself on his face without resistance, trusting to his conception of bear tactics, to save his life. Either his detailed knowledge of bear characteristics was incomplete, or this bear was an exception to the general rule, for it mauled him greatly before it was satisfied of his harmlessness.

But his only chance was to continue to appear peaceful, and he lay there until the bear left him which it did with apparent indecision as to whether it should, or no, turning back to growl and watching for signs of life, and at last following its cubs into the deep woods.

Terribly torn up, with a rib or two broken, the Indian cautiously crawled off to where he could find shelter, and be nursed into convalescence. His back was a map giving much illustration to his narrative.

This story came to me as I lay there submitting to the stings of a dozen hornets, and I kind of chuckled to think how mortified they would be if they could realize how puny their efforts were in addition to being totally irregular.

In a small way they had one advantage over the bear. It has been frequently stated that a human being in the clutch of a wild animal has little sensation of fear, and none of pain.

The Indian made the same statement.

I can assure the seeker after established facts in natural history, that in no phase of an attack by few or many hornets, do you lose personal interest in the proceedings.

Our man was rescued from the creek where he kept his head under water to dislodge the enemy from his hair, to the point of drowning himself, and was taken into camp and cared for, and where he made as close a call for life as he ever will. His face swelled until no feature was distinguishable, he was feverish to the point of delirium and it was two weeks before he could take up his work, and when he did, he was conservative to a fault, an infant might have accompanied him upon his rounds, with the cheerful consent of the most doting mother.

We worked our slow way up through the mountain Pass, and day by day we left behind us the slender thread of the mountain "Hooeyhut"[24] that was to offer the hopeful way to the expected emigrant.

It was at best but the beginning of a road, and the beginning of an emigrant road, which differs from all other kinds of roads, in its standard of practicability. The emigrant has not been educated to expect ease in transportation, and his idea of what is possible for a wagon differs widely from that of any other human being.

Given, eighteen men, a corresponding number of axes, a lesser number of shovels, and picks, and in front of them a hundred miles or so of unbroken forest, on one of the backbones of the Continent, with much broken contour, and an ultimate height of Six thousand feet to overcome, with a determination to create a pathway however crude, over which might, even though with great travail, come the men, women and children who were to make the great empires of the Pacific shore.

It demanded, First—that the eighteen should be young men to whom age had not yet brought discouragement and who deemed all things possible, and Second—that Nature should in her topography give some color to their hope.

The first we had, the second, we believed nature offered us, and we labored with a steady enthusiasm that held us up to our work.

The cohesion that held us was necessarily slight. Whether volunteers so organized would fully accept the necessary discipline of a single man was to be ascertained. Constant labor through so many hours was a preventive of discontent, but nevertheless in the evenings there was much ingenuity exercised to keep the camp cheerful.

There were debates on all subjects, singing, jesting, and interchange of experiences. And always, great fires that dispelled the gloom of the forest, and plenty of good food well cooked. Fresh meat in abundance, and Andy Burge made wise selection and the cattle that fell to his rifle from the Hudson Bay stock was tender and succulent.

Our camps were a week or more in one location, shredded cedar and fern made fragrant and luxurious beds on which to spread our blankets. As against rain, it was but a brief business to prepare, the cedar cheerfully gave its bark for our shelter.

A man with an axe, perched on another man's shoulders, cut across the bark ten feet from the ground, the edge of the axe was made a wedge, and the bark sprung open and held while the supporting man stepped backward, and a strip four or five feet wide and ten feet long was liberated by cutting across the bottom. We did not girdle any tree, a cedar twenty five or thirty feet in circumference could spare alternate strips, and ten such strips would make a shed thirty feet long perfectly waterproof, sloping from front to rear with a steep pitch.

In front, at a distance to guard against excessive heat, was a fire built along twenty or thirty feet of a huge fallen tree, at a sufficient distance from it, was a parallel log of eighteen inches in diameter, and against this would be placed tin plates with stiff cakes of leavened dough, which as they slowly baked were turned and each side secured its quantum of heat until they reached their apotheosis in a cake some two inches thick and a foot or more in diameter. One of these was no more than a meal for the individual chopper. Huge cedar bark trays with their clean white inner surfaces held any number of "collops" of beef, fried, with privilege to the epicure to broil what he desired, on switches pendant over a bed of embers. The cook cut out his trays for every meal.

Other times, beef boiled in great camp kettles, with soup galore. Potatoes and other vegetables came to camp on the pack mules.

Molasses, Sugar, Rice, Beans, peas, Dried apples (cooked with cinnamon as an adjunct) Pork, and all the other necessaries, and at meals a huge boiler of excellent coffee that placed no restraint on quantity.

Everything of the best and lots of it.

Hunger or a lack of reasonable variety, bred no discontent.

The tin plates were deep, and when we had salt pork, the grease was given to each to the limit of the holding capacity of his dish, and in the higher altitudes, mixed thick with sugar and spread on bread, made dessert.

While on Puyallup River, the cook, any time on requisition for a fish dinner, would sharpen a stake of alder and charring it in the fire to harden it, would wade out on the rapids and throw out enough hump backed salmon to make a meal for the company.

They would average three or four pounds each, and one fish, with such reasonable waste as duly considered the predatory animals that loved fish and could not so readily secure it, was about the ration for each man. When we left the river the pink flesh of the salmon no longer made part of our cuisine. I could have detailed a man to go back and bring some up, but while I guarded against satiety, yet there is no food I have ever used, that one so soon tires of, as delicious salmon and as we got to greater altitudes, the solid foods were more desired. The salmon were in immense quantities. This special kind of salmon were then running up the streams to spawn. They seemed to make it a point of honor not to turn back, and were so many that in fording, the horses trampled them under foot. A wonderful thing is the maternal instinct.

Even with the cold blooded fish, the female gives her life for expectant offspring, and presses upward untiringly and unceasingly to her spawning grounds. It is generally believed that they do not eat from the time they enter the stream from the salt water, and that they die after depositing their eggs.

Through the great ocean wastes, from unknown distances they have found their way to the waters of the streams where they were spawned, and where their descendants will follow them. Sight cannot have been of much avail in their movement. We may yet learn from them of some faculties that as yet we know nothing of.

We were traveling through a wonderful berry country and the red or blue "blueberry" growing on bushes three or four feet high were plentiful, also the Salmon berries were in full bearing. They were delicious, eaten from the bush, and that seemed to be the opinion of the bears, whose huge tracks were frequent but they feasted evidently at hours that did not interfere with us, as we seldom saw them.

Each man made a sauce pan of his own tincup and brewed a compound of berries and sugar sweetened to his own taste, which would have been perfect had not the method of its manufacture precluded all criticism of the cook. The bottoms of White River are berry paradises.

White River is a tempestuous glacial stream and is fed from Mt. Rainier. It has its regular tides, it is highest in those hours of the day when the sun melts its snow sources, and less at night and early morning when the snow melts less rapidly.

Because of the narrow strips of bottoms we were compelled to have the road make many fordings and the fords were swift and rocky. The greater number if not all could be avoided if we could afford to use our picks and shovels freely, but neither time nor money admitted of any recklessness in that direction.[25]

A Cougar in Camp

Sometimes a cougar attracted by the smell of the fresh meat would follow the packer some distance as he came up the trail but it never made an attack, and when he camped at night, a big fire kept the animal at a distance. One night I was wakened by a consciousness of danger, and just above me were two luminous eyes, gleaming in the scant firelight.

It seemed dangerous to thrust my hand further under the saddle where my head was pillowed, to reach my revolver. I could not in the darkness fully determine the position of the beast, but it seemed reasonable to assume that it was taking a great interest in my movements and not knowing what it might consider "casus belli"[26] I remained quiet.

When the great eyes shown on the other side of the camp there was still no body distinctly visible and I could not shoot. Wounded, the cougar is desperate and puts up a fight in which its own safety seems secondary to a desire to damage its antagonist.

After a time I could feel it go off into the deep woods and though I sat up for some time hoping for a clear sight and shot, I think it did not come back again. I had replenished the fire, and probably it was taken as a hint that it was not desired. I would have liked his hide, but it was too uncertain, and he might, wounded, have taken the hide off one of the men.

It was not always smooth sailing; despite of all precaution, sometimes our combinations did not accord, and we had to accept temporary discomfort. On several occasions our reliable packer missed his connection sometimes because he was delayed by the committee not having provision ready for him and other unavoidable reasons.

On one occasion we had no rations for an entire day, at another time we had nothing to eat but bacon for four days, no flour and absolutely nothing else. As a sole article of diet it is not to be commended.

Satisfactory or not, we had to keep hard at work all the time, if the men were idle in camp with this serious grievance to gloom over, there would have been a general stampede to the settlements.

All the time I was keeping silent about the news I was getting from Olympia, that there was no expectation that we would succeed in keeping the unpaid men together, and we were daily expected to abandon the work.

Each day brought its difficulties and discouragements, and each day saw them overcome, some of the men were getting letters and knew the feeling of discouragement there was about our venture.

Day by day we toiled on thro the mountains, treading where never white men trod before, through deep glens of heavy undergrowth which we cut out as we went, our campfire at night gleaming a welcome to us on our return.

Still onward, opening the dark old woods, the growth of centuries, the narrow line of light marking our progress, the entering wedge of civilization.

We were pioneers of progress, the ring of our axes was the cheering sound of a higher civilization, we were the precursors of golden harvests, the makers of new homes, and in the dim future, were visions of pleasant dwellings, blooming orchards, wind swept fields of grain, prosperous families, and the melody of childrens voices.

When we reached the Greenwater where we expected to meet the eastern party with the road completed to that point, and found they had passed in by the old Indian trail unseen by us on the road which was distant from it, and had abandoned their work, it was a discouragement hard to overcome, and we were struggling with it when relief came from an unexpected source, one of our committee men [A. W. Moore] from Olympia came out to our camp, and pushed out further to meet Capt Geo B McClelland who was exploring passes in the Cascade range for a route for the Northern Pacific R R and had entered the eastern end of the Nachess Pass.

He met him over in the Yakima valley. He had sent a reconnaissance party into the pass and had a general idea of it.

Capt McClelland had charge of the before mentioned appropriation of $20,000 for a military road and finding we had progressed so far, had agreed to accept our route and from date of interview pay for work done at the rate of two dollars a day. Further than this he could not go. He could not accept work done before that date as he could not make a post facto agreement. From this out it was plain sailing. I was in control as superintendent. I must state to the great credit of the men, that their zeal was as great when there was no prospect of any pay for their work, as when it was assured, and our relations were unchanged. We naturally all felt as if our services were receiving some recognition, and the endorsement by Capt. McClelland of our judgment in the selection of the route secured a continuance of the work next summer and an assurance that an emigrant road would be made passable as far as the appropriation would permit, and probably prepare the way for a further and more liberal appropriation, that would secure a really good mountain road that did not demand the extraordinary patience and skill of an emigrant to make use of.[27]

The passage of years or the residual effects of so many hornet stings may have muddled Robert S. Moore's memory of dates. In his 1877 account[28] he said the Allen work party left Steilacoom on June 10 instead of July 23 as reported by the *Columbian*. Ignoring the obviously incorrect date, Moore's account can be used to follow the route of the work party. He said they began cutting the wagon road from Boise Creek (near today's Enumclaw) five days after they departed Steilacoom. That would be July 28 based on the *Columbian*'s departure date for the Allen party. Meeker said in *Pioneer Reminiscences* that Allen left July 30, again pointing out the difficulty in pinning down exact dates from accounts written years later.

Kirtley and the Eastside Workers

Whitfield Kirtley left Olympia on Tuesday, July 19, with eleven men and was back in town by August 20. What he accomplished in that month is somewhat in dispute. The Longmire/Biles wagon train members said they found no real road when they reached the Naches River and had to make their own trail up the east slopes of the Cascades. Ezra Meeker pointed out that Kirtley would have had to spend at least a third of the month he was absent in transit, leaving very little time for actual work. Allen's packer, Andy J. Burge, mentioned in a 1904 letter to Meeker why so little work was done by the eastside workers. Burge stated that Kirtley's men "fell out among themselves" and that Jack Perkins came back with a black eye.[29] Burge likely heard about Kirtley's troubles during one of his many visits to Olympia and Steilacoom while picking up supplies for Allen's road builders. Perhaps he even talked with some of Kirtley's men. However, Lt. Hodges of the U.S. Army who came west over Naches Pass at the end of August said in an interview years later, that he followed the Naches River to a point where the newly constructed "emigrant road" left the river abruptly and began the ascent. Obviously some work was done on the east side for Lt. Hodges to have noticed.[30] Theodore Winthrop, who also traveled over the road in late August 1853, describes one section of the work done east of the summit in his book *The Canoe and the Saddle*.[31] Finally, James Longmire, describing the first wagon train to come over Naches Pass in 1853, said Nelson Sargent was sent ahead at the Yakima River to ascertain that the wagon road had been completed and returned with the news that they had struck the blazed trail.[32]

The *Columbian* gave one further glimpse of the road building progress that summer: "We have satisfying intelligence to communicate, that our road party working eastwardly are succeeding beyond all expectation. A note from Mr. E. J. Allen to A. W. Moore, Esq. dated August 7th, 'Away up White River,' says: 'We are going it like fire, making good time—doing much better than I expected.'"[33]

Food Shortages and Snow

Allen's November 3, 1853 letter to his brother William continues:

> How anxious all were to leave the White River Valley, and enter into that of Greenwater, where we expected to meet the other party—only to find to our bitter disappointment, when we did reach it, that they had gone on to the settlements.[34]
>
> We had ourselves been out longer than w[e] anticipated, and this disappointment seemed utterly to dishearten our men, and they were for immediately abandoning the work, and going in also. With much persuasion I induced them still to remain, and the work still slowly progressed—I keeping secret the express I received at times from the settlement, of the people having no hope of its successful accomplishment.
>
> How we resorted to all kinds of experiments to keep our minds occupied, instituting a debating society, making the grand old woods resound with denunciations of old Nicholas, the traitor Douglas, and other interesting individuals—how patriotic we all were—

"E Pluribus Unum" being the least figure of speech we condescended to use—how we sang "home songs," around our camp-fire, in the clear calm night, and each related his life experiences—how our "packer" on two several occasions missed his time, and we were entirely out of provisions, at one time having nothing to eat for a day, at another two, and on a third *nothing but bacon for four days*—no flour or anything else—and still we plodded on! On another occasion, when almost despairing, our "hearts were warmed within us" by the intelligence brought by an Indian express, that Mr. Moore, one of the committee, was coming out with provisions for us—and how, when he came, he brought none—as every one supposed we had abandoned the work.

The story, mentioned in note 3, that Allen's crew had abandoned its work has gained factual status and has often been repeated in historical literature. This error is based on James Longmire's account of his Naches Pass journey, which first appeared in print in the *Tacoma Ledger* on August 21, 1892,[35] and again in the January 1932 issue of the *Washington Historical Quarterly.*[36] Longmire said they met Burge "who had been sent out from Fort Steilacoom with supplies for the road makers, who had already given up the job for want of food which had arrived too late for them."[37] Meeker and Himes repeated this tale, adding that an Indian coming over the mountains had told the road workers there were no wagons west bound, thus sending the workers back to the settlements.[38] Allen's letter and Robert S. Moore's account showed this was incorrect. The workers did not leave the field. Allen and two-thirds of his men returned to Steilacoom on October 3. The remainder of the road workers stayed in the field a bit longer. Longmire didn't see them as Allen's men were working much farther down the trail, near the Puyallup River, when Longmire's wagon train came over Naches Pass.

Robert S. Moore explained that when the workers reached the area near present-day Greenwater, food supplies were getting low. Andy Burge, their packer, had stored half of their provisions in a hollow log near Boise Creek. He returned with some of his mules to retrieve the foodstuffs, only to discover that bears had destroyed much of the cache. Burge turned the pack mules loose to graze, and returned to Greenwater with the unwelcome news that he would have to return to Steilacoom to replace the lost supplies. Once back at Boise Creek Burge spent two days locating the wandering mules. The second night at Boise Creek, Burge dreamed of howling wolves. The dream was so real he could even hear the wolves breathing. When one wolf finally pounced on him Burge awoke, let out an unearthly yell, and sprang to his feet only to find that the wolves of his dreams were really his wandering mules who had found him in his blankets. The mules had knocked over some camp gear, startling Burge awake.

Moore stated the work crew faced a second period of short supplies as they neared the summit, going three days with only berries and a few squirrels to eat. To make their misery even more complete, a foot of snow fell on them. Fortunately it melted the next day. The work crew had nearly decided to go home when they heard Burge's braying mules coming up the trail. About this time, A. W. Moore arrived with his men but no food supplies.[39]

As to the snow, Governor Stevens in his final report to Congress stated, "In crossing [going from west to east], about September 12, in 1853, Lieutenant Hodges encountered a slight fall of snow, but it did not remain, for emigrants, crossing several weeks later found none in the pass." Allen, however, in his September 26 letter to Friend Smith said, "There has been as yet no snow upon the mountain, and will not be probably for some time." Late August and early September snowfalls in the Cascades can be very localized and are not uncommon, but they usually melt quickly. That seems to have been the case here.

Visitors in Camp

For a brief time that summer Naches Pass was a busy place. Theodore Winthrop and his Indian guide were traveling east over the mountains on their way to Salt Lake City. Lieutenant Hodges, who had been dispatched to Fort Steilacoom by Captain McClellan to obtain supplies, was working his way west accompanied by six soldiers, seventeen packers and fifty horses (various accounts give different numbers of men). McClellan made his way from the Wenas valley to the Naches summit on August 25 accompanied by Mr. J. F. Minter, an engineer, and six men. They spent the night there and then returned. He must have noticed the work Kirtley had done east of the pass since Winthrop, going over the pass behind him, made note of it. On September 6, A. W. Moore left Allen's work camp and started east.

Inevitably some of these people bumped into each other. On August 26 at 2 p.m. Theodore Winthrop going east encountered Lieutenant Hodges and his men heading west. Hodges had just passed through Allen's camp where he had stopped briefly to exchange news. Brief pleasantries were also exchanged with Winthrop and the two parties went their respective ways.[40] That night, Winthrop stumbled into Allen's camp. He described his stay quite eloquently in *The Canoe and the Saddle.* "[T]he boss of the road makers…offered me the freedom of their fireside. He called for the fatted pork, that I might be entertained republicanly…cakes of unleavened bread, hight [light] flapjacks in the vernacular, confected of flour and the saline juices of fire-ripened pork, and kneaded well with the drops of the living stream. Baked then in a frying pan, they stood…resting its edge on a planted twig, toasting crustily till crunching-time should come."[41] Winthrop continued with praise for the wonderful coffee, described a night filled will the sounds of snoring men, and noted a breakfast identical to dinner. Allen, in his comments on this meeting in the appendix of the 1913 edition of *The Canoe and the Saddle* (written fifty years after the original publication date), said he did not learn Winthrop's name at the time of his trail visit, but only years later when he read Winthrop's book. He was mistaken in this. Allen mentioned Winthrop by name in his letter of September 5, 1853, which was published in the *Columbian* on October 1. And apparently Winthrop identified Allen by name in his draft of *The Canoe and the Saddle.*[42]

On Monday, August 28, Winthrop met a member of Captain McClellan's expedition east of the summit near the little Naches River, who rode off to inform his commander that visitors were nearby. McClellan rode to Winthrop's camp that afternoon and visited briefly.[43] The Naches Trail was indeed bustling.

Allen's November 3, 1853 letter to his brother William continues:

> How, receiving a supply, we still plodded and toiled on with better hearts, upon hearing rumors that Capt. M'Cleland, (one of Gov. Stevens' company,) with his force, was coming over our road, having found no other pass from the Columbia River, up so far.[44] All these, I need not now dwell upon.
>
> Though never at a very great distance from the main body, yet during the day, with two or three others, I was generally in advance prospecting, and frequently alone—often from necessity, sometimes from choice—

"Willingly I would not be
Where life's current rolls,
Like a living sea,
Dark with human souls
I would stand where no feet trod,
Alone with thought and God."[45]

You may be sure the "eternal solitude" furnished ample food for thought and reflection. I could but reflect upon the almost incredible changes that had occurred in my fortunes, in the lapse of a brief two years; and, standing upon one of the highest spurs of the Cascade Range in rapt

"commune
High o'er the valley, where a crystal tide
Struts on its travel to the pacific wide,
With a low, merry tune."[46]

Four thousand miles from home![47] How scattered "too and fro upon the earth" were the male members of our family? I carving my way thro the Cascades—father tunneling the Alleghenies—you cleaving the southern waters, wheresoever duty calls you, "with an eye single" to *my* fortunes, rather than your own—and George, reposing in his dreamless sleep, in the sunny South—[48]

Such were some of my thoughts, thus imperfectly wreaked (I should say wrecked) upon expression, during our labors and progress

"Toilsome and slow,
Where wilds immeasurably spread,
Seeming to lengthen as we go."[49]

We still labored on—and at last the shadow in the "golden eventide" resting upon us, (as we prepared for encampment,) reaching from the last mountain we had to ascend, assured us that but a week's work remained for us, ere our task was accomplished.

Having gone over the ground between it and our camp, and marked the route, I returned with the party to our camp. I never shall forget the magnificence of the view I obtained from its highest a[l]titudes, climbed for the purpose—achieving its ascent at nightfall, and lingering till the stars came out…[50]

Having so far proceeded on our way, it was arranged that more [Andrew W. Moore][51] should proceed to the Umatillo[52] and direct the emigration this way, we having heard that agents from Oregon had been sent to warn them that our road was impassible[53]—(those who took their interested advice, were the party who suffered such incredible hardships, distressing accounts of which you have doubtless read;) and I to proceed to the settlements for provisions, of which we were alarmingly short.

The Cliff

Allen's Pittsburgh Manuscript picked up the narrative:

We had so far made a bad enough road even for an emigrant to travel over, because of our limit of time and money we were obliged to spread our labor very thinly over a great stretch of territory, and adjust it carefully to meet the conditions.

Therefore when we came to the mountain that closed up the pass as is usual with the Cascade Passes we faced an almost perpendicular ascent of probably twelve hundred feet in height, with no present means of improving it. At this point we carefully cut away the brush for a width of twelve feet from base to summit, and then guarding against the fire spreading, fired the space and cleared out the smaller growth that made it too slippery to stand on. A large tree on the summit made a "Snub" round which to take a turn with a rope, and ease the wagons hind wheels first down the slope. It was the design to leave a coil of rope for such use in case there was any prospect of emigrants that summer.

I had my own view of what could be done to make the hill passable when conditions would admit.[54]

This is the famous cliff down which the wagons were lowered and where, according to George Himes and Susan Isabel Biles Drew, oxen were killed to make rope.[55] Historian Edmond Meany had his doubts as to the accuracy of the oxen story. Accordingly he sent his research assistant to interview an ailing Van Ogle, a member of the wagon train, who adamantly denied that such an event occurred.[56] But Himes defended his story steadfastly and published a rebuttal to Meany's doubts in the January 1923 issue of the *Washington Historical Quarterly*, 78-79. Sides seem to have been taken very early as to the accuracy of this story.

This wagon wheel was found along the Naches Pass road in the 1930s and was thought to be from one of the wagons that came over the trail in 1853 or 1854. The wheel is now in the collection of the Washington State Historical Society, Tacoma. Courtesy University of Washington Libraries, Special Collections, CUR1627.

Ezra Meeker began telling Himes' version as early as 1903, repeated it often over the years, and vouched for its accuracy. At the height of the dispute in 1923, Meeker attempted to make a documentary movie about the Oregon Trail. One scene involved butchering the oxen to make rope, and lowering the wagons down the cliff. Meeker invited all Northwest pioneers to a barbeque at Naches Pass the day of the filming. He promised the pioneers nice fat beefsteaks, rather than the lean fare the slain oxen would have provided them in 1853.[57]

Allen seemed to be aware of this controversy as early as 1908 when he delicately wrote in his manuscript, copies of which he sent to Meeker and Snowden, "It was the *design* to leave a coil of rope for such use in case there was any prospect of emigrants that summer."

As the hill was timbered all the way down, it would have been easy to tie off a wagon part way down and retrieve the top rope for the next section of lowering, even if Allen had failed to leave a rope. This certainly would have been speedier than killing, butchering, and skinning oxen and then making rope from their hides. Furthermore, no account mentioned eating the butchered oxen. Even Himes, who wrote more extensively on the topic than Biles Drew, made no mention of the hungry emigrants eating the newly butchered oxen. Indeed, all that was heard from the participants was that there was a want of food, as supplies were nearly gone. Surely memories of a feast of beefsteaks would not be forgotten, even over the years. The dispute cannot be definitively settled with the available evidence. The informational sign placed at the top of the cliff many years ago by the Boy Scouts, and which is still standing, repeats Himes' story.

Allen's Pittsburgh Manuscript continues:

> Mr. Moore, the committee man referred to had left our camp with the intention of pushing forward toward Walla Walla to state to any emigrants who might consider coming the new route, the condition of the road and what was being done to further better it. Even with my optimistic opinion of the perseverance of the emigrant and of his ability to overcome obstacles, I did not feel enthusiastic about setting this task before him, in fact I did not know what he would have to meet on the eastern side. That was an open country, and a comparatively small amount of work would make it passable, but from the abandonment in so short a time I knew that little had been done on it.
>
> We worked over the divide and some little distance down Nachess river on the eastern side, and having the assurance of pay to reconcile the men to a little further delay we worked slowly back making some difficult places a little easier and making a general distribution of the additional work.

Allen's November 3, 1853 letter to his brother William continues:

> Moore returned before me, encountered M'Clelan's party, who contracted with him that, from the time he met them, those at work should have $2 per day, and I $4 which was afterwards increased to the customary mid Sept wages of the country, ($5 and $6), which, with the hope of an appropriation of remuneration for their previous labor, made the work go bravely on.

Andy Burge, in his December 8, 1904, letter to Ezra Meeker, said that on one of his supply trips he found Allen working six to eight miles east of the summit.[58] The most likely time frame for this would have been between August 29 and early September and would have put Allen's men as far east as the Little Naches River. The work crew, having been east of the summit, started working their way back over the trail. Burge and Allen continued on to Olympia to get supplies. They were in Olympia on September 15 and 16, according to the *Columbian*. As previously mentioned, many historians have stated that Allen's men withdrew from the field at this point. Allen and Robert Moore made clear that this is not true. Indeed they had a little cleanup to do north of the White River. A letter, sent to the *Columbian* probably from Steilacoom on September 15, said that before they crossed the White River for the last time, they had three miles of road to finish that they had fired (burned) on the way out. Allen left the men to do this work while he and Burge went back to get supplies. By September 25, Allen, Burge and the men were in a camp along the Puyallup River working on that portion of the road.

The following Allen letter published in the October 1, 1853, issue of the *Columbian* tells more of the progress of the road workers.

> PUGET SOUND EMIGRANT ROAD,
> September 5th, 1853.
> FRIEND McELROY: [editor of the *Columbian*]
>
> You will by this time have almost forgotten my promise to write you a line occasionally, telling you of our progress upon the road. I have managed now and then to have you learn our where-abouts, but have never written directly, for this obvious reason; knowing that our party have been detained upon the road much longer than was anticipated, in consequence of having to cut farther than was expected, I was fearful that we would not be through in season, and so refrained from writing lest I should say more than after events would

justify. From where I write this morning, the shadow from the mountain side falls upon my paper, and as the terminus of our work lies some few miles from its base, I feel confident that this end of our route will be completely finished. We will probably turn back to White river within a week. Before crossing White river for the last time, we have some three miles of road to finish, in which we put fire as we came out. This will bring us out upon Tanalquot prairie, and from there we strike the old cut road leading out towards "Muck."

I am much better satisfied with the route than when I started, though as you may recollect I was not a little sanguine then. It will make an excellent road, and all who travel over it speak highly of it. I had the pleasure a week or so ago to see Lieutenant Hodges, who has gone in to Steilacoom to pack out provisions for Capt. McClellan, who is now camped near the Yakima. The lieutenant speaks in the highest terms of our road, and of its superiority over all other passes yet seen by them. We have but one really bad hill, down the mountain side, which is with our limited time and means for improving, without doubt a bad place, but can be made with but little outlay very good. It is not, however, to be compared with "Laurel Hill" upon the other route, and it rests upon its own merits alone, as it has not like Laurel Hill, infinitude of others to back it.[59] I only regret that we cannot remain long enough to make it as complete as we would wish, and so made the whole road so excellent, that the idea of crossing the Cascade range would be no longer as heretofore a bug bear, and a terror to the toil worn emigrant. The grass also is much better than was anticipated. The prairies upon the summit are now waving with luxuriant growth of grass, amid which I have noticed timothy and clover. There is no laurel upon the whole route, neither (as far as I can ascertain) any poisonous plant to injure cattle, and this, to those who have crossed the southern Cascade road and have lost animals thereby, will be an item of no small importance. On the whole, the people of Washington Territory will have reason to be proud of their "people's road."

By some error the list of men working upon the road was mislaid before reaching you. I copy it again for you the more readily now, because of their unwavering perseverance in adhering to the road despite of all discouragements and disappointments, coming out as they did with the view of remaining one month, and the prospect now looming out large of consuming two.

(The names we gave last week.)

Meeker supplied a complete list of the road workers.[60] He obtained the list of Allen's men from various issues of the Olympia newspaper.[61] No record has been found as to when or how Allen and Kirtley were appointed to lead their respective work parties but having been road viewers, they were the obvious choices.

The spellings of a few names in Meeker's book differed from those published in the *Columbian*. As Meeker knew most of these men personally, he presumably corrected errors in the newspaper accounts. The authors have added missing first names and middle initials where known and supplied the different newspaper spellings. Meeker's spellings come first where there are two names.

West slope workers
E[dward] J. Allen
A[ndrew]. J. Burge
Thomas Dixon
Ephram Allen [Allyn]
James Henry Allen [Allyn]
George Githers
John Walker
John H. Mills
R[obert] S. M[o]ore
R[obert] Foreman [Forn]
Ed[ward] Crofts [Crafts]
Jas. Boise
Robert Patterson [Pattison]
Edward Miller
Edward [P.] Wallace
Lewis [K.] Wallace
Jas. [Joseph] R. Smith
John Burrow
Ja[me]s. Mix
Jesse Boyes

East slope workers
Whitfield Kirtley
Edwin Marsh
Nelson Sargent
Paul Ruddell
Edward Miller
J. W. Fouts
John L. Perkins
Isaac M. Brown
James Alverson
Nathaniel G. Stewart
William Carpenter
Mr. Clyne

In 1887 Robert S. Moore left an account of these events. While written thirty-four years after the events it chronicles, it added much to the story. Moore said J. E. Williams and a Mr. O'Connell left the crew to obtain horses in the Yakima Valley to plow land on claims they had taken on Connell's Prairie.[62] Neither Meeker nor the contemporary newspaper accounts mentioned them. Notes in the Allen materials at the University of Pittsburgh library stated that Joseph R. Smith was better known as Russell J. Smith.

Allen's letter resumes:

The greater part of these have taken claims upon the White river prairies, where I have seen the most desirable farming lands probably in Washington Territory, combining good soil with a reasonable vicinity to good markets upon the Sound. The greater portion of the men working upon the road depend upon their labor for their subsistence, and have now the prospect before them of the rainy season coming in, and no provision made by them for winter. It is not asking too much of our citizens, to request through your paper that those who delayed giving before, because doubtful that the road would be completed, now to give liberally to enable the men who have worked upon the road so faithfully, to receive at least a small portion of their wages immediately, so that they may be able to await Legislative or other action for the forthcoming of the balance.

To-morrow [September 6] Mr. Moore sets out for Walla Walla, to endeavor, if not too late to turn the emigration, to bring in the first train which has ever entered Washington Territory by the People's Road. I should be but little surprised if Governor Stevens made his "entreo" by the new route. Mr. Winthrop passed some ten days ago on his way to the States via Salt Lake.[August 26]

I have some curious specimens to show you upon my return: some copper ore, pure brimstone, and something almost coal—the two former in large quantities—the latter with considerable "klonass"[63] attached to it.

I have a doubtful way of sending these few lines to you, but hope they will come safely to hand, and so remain

Yours Respectfully
EDWARD J. ALLEN.

The First Emigrant Arrives

Allen's November 3, 1853 letter to his brother William continues:

We were camped one day, on Puyallop River; it was raining dismally—one of those determined kind of rains that augers no cessation. Nothing of so trifling a nature had hitherto prevented us from working, but the day itself was cold and dreary, a mist seeming to hang over the earth, and a denser one over our hearts—one of those days you have so often in Pittsburg, proof against epidemics, but death on bronchial or consumptive complaints. I had strayed from camp, partly occupied in examining a little spur or hill, as to the comparative advantages of grading or finding a pass to avoid it—and partly given up to reverie, indulging, as I walked along, in the pleasing fancy that the mysterious whispering of the tall trees above me were "home voices"

"Oh, dear deceit!
Whispering ever that fond hearts have found us
And in the distant desert breathing round us,
Their voices sweet."[64]

Thro the opening I discerned advancing toward me—"a solitary horseman—"

"All dreary was his semblance,
And little was his pride;
His one foot in the stirrup hung,
The other waved beside;
His hat was hanging o'er his eye,
Simple was his array;
A sorrier, sadder man than he
Rode never a rainy day."[65]

—The most forlorn looking individual I had ever seen—and I have seen some distressing specimens in my time, beside affording a pretty good imitation myself, on sundry occasions. He was as haggard, despairing, wistful, and lean looking "unfortunate" as—well, as that poor fellow who enveigled me out of my last herring![66] I do not think, seriously, he had strength enough left to kick a gooseberry a yard before him. His story was soon told. He [James Aiken][67] was one of the advance emigration, who had ridden on to solicit provisions for his starving comrades left behind.

We were not very flush of provisions, ourselves; but half an hour after one of our party was dispatched with 300 pounds of flour to meet them with orders to give to all freely—and then, giving our pioneer a dinner, (all of our party gathering around, to witness his performance,) and a fresh horse, I rode in with him, riding hard all day and reaching Usiqually [Nisqually] after midnight. No one charged him anything; he had free passage to Olympia, whither he wished to go. I procured a fresh supply of provisions for our party, and started back next day; met Moore (who was clerk of the county,) coming in; reached the camp a few days afterwards.

Allen's Pittsburgh manuscript supplies additional details:

{While on the Puyallup near to the settlements, we were overtaken by an emigrant the first man on the new road, who had come in advance to procure provisions, as his people were very short.

{We had not at this time much of a supply as we had allowed ourselves to run low as we got nearer home, but within half an hour one of our party was on the road eastward with three hundred pounds of flour to distribute to those who were in need, and then we gave our visitor a good dinner of viands that he had not seen for some time, and I then rode in with him to the settlement, where he received a generous welcome and was forwarded to Olympia where he desired to go.

{I took out a little further supply of provisions to make the amount necessary after draining ourselves to supply the emigrants and with a little further work we packed up and came in having been out on this arduous work one hundred and forty days.

{Nearly five months without any compensation except for a little time guaranteed by Capt McClelland.

{A pretty good proof of the patriotic feeling in the territory when these people gave their time and labor for such a period and the community contributed the provisions and the money. Such a community is a good one to live in.}[68]

Robert S. Moore's 1887 account differed from Allen's only in regard to how Aiken encountered the work camp:

About the fourth or fifth of October, while we were camped at the Puyallup river, opposite the present Van Ogle place, at midnight there came across the river a loud "Hello." We were somewhat startled for we did not know of any whites being in our rear. When we answered we were told there were seventy wagons with emigrants behind and many were short of provisions. Having known the pangs of hunger ourselves, we told the man to come across to camp. He was Mr. Akins. He said he was one of James Byle's company. All hands turned out and shook hands with the first emigrant who had crossed on the Naches Pass wagon road. As soon as it was light enough, we packed up all our provisions and sent Andrew J. Burge forward with all dispatch. We confiscated an Indian Potato patch and with dog salmon we finished the road.[69]

After hearing Aiken's story and sending Andy Burge up the trail with the last of his supplies, Allen rode to the Nisqually ferry with his visitor. From there Aiken, with a note from Allen, continued on to Olympia to secure help while Allen went to Steilacoom to replace the supplies he had sent forward to the Longmire/Biles company. Allen's note was printed in the *Columbian.*

STEILACOOM, Sept. 26th, 1853.
FRIEND SMITH:[70]

This will be handed to you by Mr. Aiken, the first emigrant who has crossed upon the "Puget Sound Emigrant Road." He left a train of 35 wagons upon the Wenas River, and reports 60 in all as having crossed the Columbia. Some of them are out of flour, and he brings in a letter addressed to M. T. Simmons, praying the Washingtonians to send out relief to them. I have sent out from our camp on the Puyallup 300 lbs. flour, and have come to Steilacoom to pack out more for our own use—The train is under the guidance of Nelson Sargent.

Capt. McClellan's company discovered gold upon the Yenass. Some of his men told Mr. Moore that with a "long tom" they could wash out $4 per day.

CAPT. McCLELLAN REPORTS NO PASS BETWEEN OUR ROUTE AND VANCOUVER THAT IS PRACTICABLE FOR A WAGON ROAD, and has in part accepted our work, subject to the approval of Gov. Stevens.

There has been as yet no snow upon the mountain, and will not be probably for some time. There is only about ten miles of the whole route upon which snow lays to any depth even in mid-winter.

The first opening of the road through our heavy fir timber, however great the amount of labor expended, must necessarily leave it a rough road; but this year's travel, and the work which next year will be given to it as a military road, will probably make it the best road of its length in the two territories.

We will be through this week, having completed the western portion of the road. I leave to-day with pack animals, and will be at the camp on Puyallup to-night.
Yours,
EDWARD J. ALLEN.

The *Columbian* announced Aiken's arrival.

Mr. Aikin, the first immigrant by the PEOPLE'S ROAD, reached this place on Monday last [September 26], via Steilacoom. He left his train of 35 wagons on the Wenas river under the charge of Mr. N. Sargent, and came in to procure flour—his party being nearly, if not entirely without. Mr. Allen at once dispatched 300 lbs. from the camp of the workmen on the road, near Puyallup river. By this commendable promptness on the part of Mr. A., the train will be enabled to hold out until the ample supplies sent out from this place reaches them. So soon as it was known in this place that the immigrants were short of provisions, Col. Simmons, and others, took active measures to send out relief, and Mr. Aikin was enabled to leave for his train the next morning with 1000 lbs. of flour, onions, etc. The following persons accompanied him from this place to render assistance, viz: B. Gordon, T. Bush, J. Kindred, M. Jones. On the day previous, O. Cushman and W. Sargent left this place to meet the immigration and lend their assistance in getting through. Mr. Aikin reports sixty or seventy wagons as certainly coming the new route, and having crossed the Columbia river for that purpose. They are determined to come in through the new road at all hazards, rather than follow the old one into Oregon.[71]

Rescue

Burge crossed the Puyallup River and traveled as rapidly as possible into the mountains with Allen's three hundred pounds of flour. He found the wagon train a few miles west of today's Government Meadow. While the Longmire/Biles men were lowering the wagons down the infamous cliff with ropes, Mrs. James Longmire, Mrs. Erastus Light, and their children went ahead on the trail. As they walked along they ran into Andy Burge. They were more than pleased by his rude welcome. "Good God almighty, woman, where did you come from?"[72] Burge continued on to meet the wagons, gave them the flour, and quickly returned to Steilacoom to get aid.

James Longmire gave a puzzling account of his meeting with Burge, saying he (Burge) tried to persuade the emigrants to go back to where there was water and grass for the stock "telling us it was impossible for us to make our way over the country before us. Failing to convince us of this, he set to work to distribute his supplies among us, and returned to Fort Steilacoom, blazing trees as he went and leaving notes tacked up, giving what encouragement he could, and preparing us in a measure for what was before us. For instance, he said, 'The road is a shade better.' A little farther, 'A shade worse.' Then again, 'A shade better.' And so on till we were over the bad roads."[73]

Longmire either misremembered after the lapse of years or embellished a story that needed no such help. Burge had been over the western section of the trail many times in the previous two months. He knew it was passable for stock and most likely wagons. Indeed, a steady stream of traffic had been over the trail during the summer. Dozens of horses and mules had made the mountain crossing with Indians, road builders, the military, and even the odd tourist such as Winthrop. Burge knew the wagons at that moment were being lowered down the worst remaining obstacle on the trail. He also knew that grass grew at Bear Prairie, just a few miles ahead. Burge may have encouraged the party to rest and graze their stock at the meadows west of the summit before continuing on and he may have tacked notes to the trees, but he would have not reasonably said the trail was impassible. No other contemporary account mentioned this supposed discussion between Burge and Longmire.

Another possibility is that Longmire said no such thing. The Longmire narrative was actually written by Mrs. Lou Palmer and was based on a series of interviews she had with Longmire at Longmire Hot Springs a few years before he died. (Mrs. Palmer's grandfather was apparently a member of the 1853 wagon train.) Mrs. Palmer's narrative was first published in the *Tacoma Ledger* on August 21, 1892. She wrote using James Longmire's voice, but the words were hers. She may have been inaccurate in taking notes and converting them into text.

According to both Light[74] and Longmire, the Simmons-sponsored relief party with its thousand pounds of food reached the wagon train at the foot of Mount LeTete, near the present-day village of Greenwater. This would have been no more than two days after Burge delivered his supplies. Perhaps this is also where John Parker and Quincy Brooks met the immigrants.

Allen's November 3, 1853 letter to his brother William continues:

Having been engaged on this arduous voluntary duty, 140 days,[75] we packed up, returned, and reviewing our work as we passed along, left behind us the hills, and emerging on the last ridge, looked out upon the plains, on Usiqually [Nisqually] Prairies, where,

> Smoother vales extend
> Immense horizons—vales succeed,
> Far as the eye can see, without an end.[76]

Beyond lay the sounding bay, with embryo Venices and New Yorks already rising on its shores—a space we soon rode over, and dashing into Stillacoom, cheered as we went, for the success and full completion, under so many disheartening discouragements, of the Cascade and Puget Sound Emigrant Road.

The whole town—I may say Territory—was alive with excitement and enthusiasm, on our arrival, which was increased by the arrival of the first train of the emigration of '53, which proved beyond a doubt the practicability of the road, the assertions of our parent Territory, Oregon, to the contrary notwithstanding. Henceforth for all future years, the new comers will be spared the tedious and hazardous winter journey down the Columbia, and then over land to Puget Sound.

The papers you have received, containing the account of the successive public meetings held in celebration of its completion—the eulogistic resolutions passed, votes of thanks and commendations from all sections of the Territory, will spare my modesty their repetition here. I was truly, for the nonce, "a distinguished individual"—the guest of the Territory. I could not say, like St. Paul, "no man was at charges for me," for (to the sacrifice of my self-dependence) I was not suffered to pay for any thing received. Not the least gratifying of all those testimonials, evincing that generally I had become a "popular institution," was the proud satisfaction of having been able to assist the starving—the grateful acknowledgment from the emigrants we had relieved, and more than all, the public testimonial of esteem and friendship from the faithful compeers and comrades, (who, under so many discouragements, had so nobly stood by me,) on the eve of our parting, each to our several destinies on the broad current of life.

Nelson Sargent, according to the August 20 issue of the *Columbian*, had "gone forward [over Naches Pass] to meet the immigration and conduct them in." Allen materials at the University of Pittsburgh said, "As nearly as can be determined he (Sargent) must have left the road party on the east side of the Cascade mountains about August 8th, and arrived at the camp [in the Blue Mountains] referred to, 225 miles eastward, five days later." Thus Sargent averaged forty-five miles a day on horseback. He found the Longmire/Biles wagon train and his father's family in the Blue Mountains of eastern Oregon. There he told them of the new wagon road under construction over Naches Pass that led directly to Puget Sound. The weary emigrants agreed to become the first to travel over the "People's Road." The *Columbian* on September 24, 1853 noted, "An Indian reported Mr. Sargent to be on the way through with a train of sixty wagons. We hope this may prove true."[77]

The *Columbian* of October 15, 1853 reported:

Our attentive friend, Capt. E. J. ALLEN informs us that eleven wagons have reached the shores of Puget Sound over the Washington Territory Emigrant road. The following are the names of the immigrants so far as learned: John W. Lane and wife, Samuel Ray, W. Ray, Henry Mitchell, H. Rockenfield, James Barr, J. A. Sperry, J. Longmire, wife and five children, William Clafland, Evan Watts, E. A. Light, wife and child, J. J. Ragan, William Kincaid, wife and six children, William McCreary, Isaac Woolery, wife and three children, Peter Judson, wife and two children, G. Miller and John Neson—total 46.

The immigrants bring 62 head of working cattle, 20 cows, and 7 American mares—all in very good order after so long a journey. The immigrants are well pleased at their safe arrival and represent the road as much better than they expected to find it.

We are happy to learn that the supplies forwarded to their relief, reached them in time to prevent suffering. The 300 pounds of flour dispatched by Mr. Allen reached the train just as the last pound they had was being cooked. They are deeply impressed with the liberality displayed toward them by the citizens of the territory, and are grateful therefore.

In early October 1853 the wagon train arrived at Puget Sound. Interestingly, the participants did not all agree as to their arrival date, giving ample testimony to how memories fade over the years. James Longmire and Erastus Light said they reached Clover Creek on October 8 and came into Steilacoom on October 9. James' son, David Longmire, said they arrived at Clover Creek on October 12. Van Ogle did not give an exact date for arriving, but he did supply some dates. He suggested they reached Naches Pass September 28, and after the wagons were all roped down "Summit Hill" six miles west of the summit, it took them seven more days to reach Boise Creek and three more days from there to Clover Creek, suggesting an October 10 arrival. Himes put the wagon train on the summit on October 1 and agreed with David Longmire that they arrived at Clover Creek October 12.

Honors for the Road Workers

The honors Allen wrote his brother about came quickly. The *Columbian* of October 22, 1853 reported:

MR. EDITOR:---The people of Steilacoom had the great pleasure yesterday, of receiving and entertaining, a party of immigrants just arrived by the new road opened by Capt. Allen and party. They are all in good health, and have a respectable amount of stock, in good condition. They are somewhat disappointed in the quality of the greater portion of the soil, but at the same time determined to become permanent settlers and contented men.

The *Columbian* also paid tribute to Nelson Sargent. "We see that Mr. Nelson Sarjent has reached home again. He went out as far as Burnt river to meet his father and family, who were on the way to the territory. To Mr. N. S. belongs the honor of having piloted the pioneer train of wagons over the new route."[78]

The members of the wagon train were not so kind in their treatment of Sargent. Meeker reported much venom was directed at him as they struggled through the Cascades. Some felt that Sargent had misrepresented the quality of the "People's Road" to them. They apparently expected better. Meeker felt Sargent did not mean to deceive, but was simply overly optimistic as to what the road workers would accomplish from the time he left them until he returned a month later guiding the wagon train. In a letter to Ezra Meeker, looking back at the project fifty-two years later, Allen recalled that the road was "not sandpapered."[79]

Allen received tributes also. The *Columbian* on October 8, 1853 printed the following:

Tribute to Edward J. Allen, Esq.

It is with unfeigned pleasure that we give place to the following card handed us by the men who were engaged upon the western extremity of the Cascade Road, and the compliment paid Mr. Allen is no less sincere than deserved.

Steilacoom, October 3rd, 1853.

We the undersigned laborers on the western extremity of the WASHINGTON TERRITORY EMIGRANT ROAD, hereby tender to Capt. E. J. Allen our most sincere thanks for his gentlemanly conduct, manly exertions for our comfort, and perseverance in the prosecution and completion of an enterprise which must ever reflect credit upon its projectors—among whom, we are happy to say that Mr. Allen has not a little distinguished himself by his zeal, energy and unremitting efforts, under circumstances very discouraging attending so long a trip, and the difficulty of sustaining friendly feelings and unity among men that had no provision made for the winter, and consuming more time than was expected when they went out.

E.S. Moore,
Jesse Boyes,
James H. Allyn,
John H. Mills,
Robert Pattison,
Ephriam Allerd,
Edward P. Wallace,
Edward Miller,
Lewis K. Wallace[,]
Robert Forn,
Edward Crafts[,]
James Mix,
Thomas Dixon[,]
John Walker,
Joseph R. Smith.

The men who were still at work on the road and thus not available to sign the card were: A. C. Berge, Jas. Henry Allyn, George Githers, John Walker, R. S. Moore, John Barrow, Jas. Meeks.

Apparently, Allen left one-third of his workers to finish up along the Puyallup River and he came into Steilacoom with the rest of his crew on October 3, again contradicting both Meeker and Longmire. The newspaper added:

The citizens of Washington territory are requested to meet in Olympia, *en masse*, on SATURDAY, October 15th, 1853, to celebrate the arrival of the FIRST immigrant train over the People's road. "Keep the ball in motion," and let every one come and join in giving them such a welcome as only Washington Territory can give. The "Cascade Road Committee" desire all those who subscribed to this road to come forward with the amount of their subscriptions, at an early day as possible, so that the committee may be enabled to make a final report to the mass meeting, to be held in Olympia on the 15th inst. Those who have not heretofore subscribed anything, ought now to do so. The road is now a reality—A FIXED FACT—THIRTY FIVE WAGONS HAVE PASSED OVER IT—and you are as much benefitted by it as anyone else, and now is the time to reward the untiring perseverance and industry of those men who have accomplished the GREAT WORK.[80]

The mass meeting was held at 1 p.m. in the hall over the store of Parker, Colter & Co.

One week later, on October 15, the *Columbian* reported that "Mr. JOHN M. CHAPMAN, proprietor of Steilacoom city, has given a town lot to each of the men who worked on the western section of the "Washington Territory Emigrant Road."[81] Lafayette Balch offered a similar reward to the road workers in his part of Steilacoom. No such reward was offered to Whitfield Kirtley and his men.

Allen's November 3, 1853 letter to his brother William comes to an end:

Some of these men had taken up claims in the beautiful White River valley we had made accessible, and were going to reside on it. To each and all were presented, by the large-hearted Capt. Sebastian [Lafayette] Balch, (proprietor,) a lot in Stillacoom, as also one large and very valuable one to me. Gratifying as all this was, together with other flattering incidents, the offers from divers energetic capitalists to embark in various apparently profitable undertakings, (all of which I was compelled to decline till I heard from you,) there was still a shadow resting upon me—my own personal affairs had not prospered as well as the projects undertaken "pro bono publico."

You recollect, Ensign being a ship carpenter, I had sent him on to Wallah-Wallah to "institute" a ferry there. It was a large undertaking, so far from all conveniences, cost a good deal, and was not completed in time to do much business, but it will be ready for next year. This was not all; before starting out to cut the road, I had left some men on my claim to get out timber. There had been a glut in the California market, and the demand had ceased temporarily. Squared timber is so generally a cash article; and at the same time the land was being cleared—that I continued clearing, writing meanwhile to San Francisco to enquire the selling price, delivered there, and cost of chartering a vessel to come for it. One day, when in town procuring fresh provisions for our party, I received a favorable answer.

It was a severe blow, in the midst of the ovations tendered me on my return from the completion of the road, to hear that a fire had broken out on my place, and consumed my house, and lumber to an amount that would have sold for $2,000.[82]

It required considerably more than the philosophy of Mark Tapley,[83] to come "out jolly" under such intensely aggravated circumstances. I had worked hard, had "done a good work," needed rest, and my home, humble as it was, had loomed up pleasantly in the darker hours of discouragement on the road as a retreat. I conned over Longfellow's verse (wonder if he ever had his house burned down?)

"Let us then be up and doing,
With a heart for every fate—
Still achieving, still pursuing!
Learn to labor and to wait."[84]

But I could hardly flatter myself I derived much consolation from it; seemed "cold comfort"—and I fancied the author had no conception of such a distressing case as mine, and rattled them off as a mere exercise of his poetic faculty, instead of conveying a moral.

My mental barometer, however, generally indicates "fair," and a reaction soon was the result. I thought of the great boon of health, granted me; the life of free and unrestrained action before me, youth to achieve it; of you all at home, and a voice seemed to say

"Far as thou art,
Our spirits wander with thee, and are here—
In peril or in pleasure ever near—
Twined with thy heart!"[85]

I shall rebuild my house again, and as soon as completed reside in it again; engage in nothing for a time, wait awhile—not with the vague hope of that sanguine individual, Mr. Micawber,[86] of "something turning up," but watch the "coming events"—and pitch in at the precise time that in my judgement "leads to fortune."

Every one is anxiously waiting the arrival of Gov. Stevens, and all his detached parties of exploration, and no one more impatiently than myself. If he finds a pass northward of the South Pass, and at a lower attitude, it will redound to his fame. If Captain M'Cleland (whose mission it was to find the best pass thro the "Cascades,") finds none better than ours, which is called after its Indian name—the "Narches," Northward, it would not only ensure us pay for the work done previous to his crossing, sooner than perhaps our own Legislature could have a special act passed for the purposes, but it would increase the chances of the Governor (who alone has the power,) of selecting it as the best road to be finished for a "Military Road," for which the appropriation of $20,000 was made by the General Government, thro Gen Lane's influence, and 5,000 of which has already been uselessly expended in two futile attempts to accomplish it.

As I have got to be somewhat experienced in road-making, (the "census man" was puzzled at my frank confession of having no special trade or profession—and knowing me only in connection with the "Cascade Road," put me down as a civil engineer)[87] my first object would be to obtain the contract from the Governor to finish it; as all my old compatriots would flock to me, I have no doubt I could make it profitable, and give satisfaction.

I am residing, (pending the redivivus of my house,) "in spots" generally spending my time mostly on my place, which is becoming more valuable every day and especially since the completion of the "People's Road," which has given an impetus to the whole territory. Indeed my home is becoming very dear to me, alone as I am; desolate and drear as it looked, when I first came to it after so long an absence, it was still beautiful. It was a beautiful day, and the "Time of the singing of the birds" had not passed, tho with you it was no doubt stern winter; the woods were thronged with the feathered songsters, and I felt as thankful as the old ballad-maker, Robin Hood, when after his restless residence a year at the court, he sloped for his forest home.

"It was upon a merry morn
That he came to the green wood—
And the merry notes of the little birds
They were to his heart good."[88]

—I must close. I write this at a neighbor's. I have been roaming all over the place to the other side, which is bounded also by Ells' Inlet, a sheet of water wider at my place than at Budd's Inlet. I would give half I possess to have you all see it with me, at this time, and from the point I occupy. The day is fading—
Fervently yours,
Edward Jay Allen

The Mitchell Wagon Train

A second wagon train that came over Allen's road in November has received little notice in the historical record. The November 19, 1853, issue of the *Columbian* suggested its story was at least as interesting as that of the Longmire/Biles wagon train. "A party of immigrants reached Steilacoom City a few days since, having come through the new route. Owing to the lateness of the season, they were caught in the snow before getting out of the mountains and suffered severely. Fortunately assistance sent out from Fort Steilacoom reached them in time to prevent actual starvation."[89]

This party included William H. Mitchell who came from Kenosha, Wisconsin, with friends also heading west. The Mitchell train got a late start, crossing the Missouri River June 3, and reaching Fort Kearney on July 4. When they came over the Blue Mountains, winter was approaching. At Fort Walla Walla the wagon train divided, with part going down the Columbia River and part going north through Naches Pass.

William Mitchell described the trip over the pass:

When we arrived at the foot of the mountains two men were sent forward to see if we could get over and they returned, reporting too much snow[90] for the wagons, so it was decided to leave our outfits at a Catholic mission[91] that was there and proceed on foot, letting the women ride whatever there was to ride. Mr. Wooden and myself were the first to start over the pass and we found the way not nearly as bad as represented and by taking advantage of cut-offs, we made very good progress and without misadventure until the last night in the mountains we became separated. Mr. Wooden took what he thought to be a cut-off while I stayed in the trail, and, at night, as he did not rejoin me I called him but received no answer. . . . [I]n the morning. . . it was raining hard, so after stretching a piece of canvas over some brush to keep the rain off, I built a fire and was cooking the last of my store of rice when Mr. Wooden came into my camp. We ate the rice that I had prepared and started on our way again, and that afternoon met a Mr. Connell, who gave us a little flour which we cooked on the end of sticks and ate. We took supper at his house and then, after hiring horses from some friendly Indians, pushed on to Fort Steilacoom.[92]

A complete roster of this wagon train has not been located, but Mitchell supplied a few names. (George Himes named 148 members of the Longmire/Biles train, a list Meeker published in *Pioneer Reminiscences.* A few years later David Longmire published a more comprehensive list.) Mr. Holmes and his family settled near Olympia. Mr. Wooden started the first tannery in Seattle. Mr. Schock and Mr. Livingston with his two daughters settled near Seattle. One of the daughters married Will H. Brannon and located on the White River where they were killed during the Indian war. Mr. Johns and his eight children, one of whom Mitchell eventually married, built on the White River and apparently survived the Indian wars. Bird Wright and his two brothers and families settled in the Puyallup valley. (The Wrights also appeared in the list of the Longmire/Biles train. Perhaps they dropped behind and joined the Mitchell train. Dorothy Winston placed them in the Longmire train, describing Bird Wright family happenings that seem right out of Longmire's reminiscences, destroying her case at the end by announcing that the Wrights arrived at Fort Steilacoom November 2, 1853.)[93]

The emigrants had arrived. Allen's road was completed in time to serve its purpose that fall, bringing a contingent of much-needed settlers into Puget Sound. Many more would come the next season via the People's Road.

The Captain and the Colonel

Roommates Reunited

Allen rebuilt his burned-out cabin in the fall. That winter he had two notable long-term guests—Captain George B. McClellan and George Gibbs.[1] McClellan became nationally known as a Civil War general and as the Democratic Party's 1864 Presidential nominee. Gibbs, less well known, was a New York lawyer who crossed the plains as a civilian in 1849 with the Oregon Mounted Riflemen, the first U.S. military expedition to travel the full length of the Oregon Trail. He became a student of the various Northwest Indian languages and was employed as an ethnologist and geologist by McClellan in July 1853. In 1854 he accompanied Governor Stevens on his Indian treaty-making journey through western Washington.[2]

Years later, a Pittsburgh newspaper correspondent interviewed Allen.

Col. Allen is very fond of wood fire.[3] The flicker of the flames seem to stimulate his recollections, and it was before a wood fire in his library, the only light being that of the fire itself, that he told me several interesting anecdotes.

One of them related to an early experience of his with George B. McClellan, who was at that time a brevet captain[4] in the United States Army, and one to a later experience when this brevet captain had become major-general in command of the Army of the Potomac.

'I was living in the early fifties,' said Col. Allen, 'in a cabin which I had built in Olympia, in the state of Washington. I was very glad to see any new face and always tried to make a visitor comfortable. There came one day to my cabin a young man who was of most attractive personality. His name was George B. McClellan, and he was a brevet captain. I found McClellan a most amiable and engaging character. I invited him to share my hospitality for as a long a time as he wanted to. He accepted my invitation in the spirit in which I gave it, and he lived with me in my cabin for several months. A strong attachment was established between us, and, although after his departure I did not see him again for years, yet I was loyal to the friendship which was at that time established.

I had built the emigrant road over the Natchess pass. It was a very formidable undertaking and of a kind which none but a young man who dares to venture anything would have attempted.

Capt. McClellan became familiar with this emigrant road and he often spoke to me in terms of high commendation of it.

Well, when I learned that George B. McClellan had been appointed commander of the Army of the Potomac I determined to renew my acquaintance with him. I wanted to call on him to pay my respects and give him my congratulations. But when I went into his office what should I see but a bewildering array of uniforms there. Some of the men were foreigners, and I afterward learned that they were distinguished Frenchmen who were volunteers on his staff.

These men did not look upon me very encouragingly. The orderly who came to me and who saw that I was in civilian dress said, 'Gen. McClellan is very busy and cannot possibly see you.' Thereupon I wrote upon my card these three words: 'Clahyam six tilicum.' Those words are Chinook and in English were translated as 'How are you friend?' I thought they would recall those days in the early fifties if McClellan saw them.

The orderly took the card in, after some urging on my part. In an instant out came McClellan, smiling, holding out both hands with utmost cordiality and interest. He took me immediately past the uniformed splendors into his office.

After a brief chat he said to me, 'Allen I want you here. You must take right hold in my commissary department. I can't tell you how good it seems to me to see you.' Before I departed he wrote a letter to President Lincoln direct, urging my appointment in the commissary department.[5]

Allen learned Chinook jargon quite early in his stay in the Northwest and was often called on to translate during encounters with Northwest Indians. He said, "I knew George Gibbs well. I was with him while he was compiling his Chinook jargon dictionary.... For my coming volume, 'The Oregon Trail,' I have in manuscript perhaps the fullest vocabulary of the jargon yet offered. It is compiled from all the vocabularies to which I could get access, together with the pronunciations as I knew them."[6] McClellan's knowledge of the Chinook jargon was no doubt enhanced by George Gibbs, his aide and ethnologist, who accompanied him on his exploration of Washington Territory in 1853. Gibbs published his dictionary in 1863. Allen's unpublished dictionary may be found in the Hervey Allen collection at the Hillman Library, University of Pittsburgh.

Plans to Survey the Cascades

Washington Territory was created by the second session of the 32nd Congress (1852-53). On March 2, 1853, the newly elected President of the United States, Franklin Pierce, awarded the governorship of this new territory to one of his supporters, Isaac Ingalls Stevens. The Congress also directed that several surveys be made to locate a route across the United States for the future Pacific Coast railroad. Stevens was assigned to survey the northern route. Born and raised in Massachusetts, Stevens graduated from West Point at the head of his class, and was serving as a brevet major in the Army Corps of Engineers when he received his appointment as governor. He spent most of 1853 slowly surveying the railroad route across the northern prairies, arriving in Olympia in November to take up his gubernatorial duties.

Anticipating that he would not reach the Cascade Mountains before winter, Stevens delegated one important task. As mentioned in previous chapters, Congress had appropriated funds for the construction of a military road to run from Fort Steilacoom on Puget Sound, east across the Cascade Mountains to Fort Walla Walla on the Columbia River. The 1852 Congress would not appropriate money to build an emigrant road across the Cascades. However, Oregon territorial delegate Joseph Lane secured a $20,000 appropriation in late 1852 by simply calling it a military road. Allen later ranted about the hypocrisy of a congress that was, as he wrote:

...So opposed to appropriations for internal improvements, that no measure could have been passed for an emigrant road; [but] such a bill was passed for a military road from Fort Steilacoom, an encampment on the prairie, to Walla Walla, a Hudson's Bay fort on the Columbia, with an appropriation that would not have even paid for an engineering survey. Naming two forts made it seem plausible, whether the sapient M Cs [members of Congress] who voted for the bill thought both named forts were U S, and were willing to make such an appropriation and 'save their face' by calling it an appropriation for military uses, I cannot say, but the appropriation was made."[7]

Hypocrisy or not, Stevens delegated this construction job to U.S. Army Captain George B. McClellan. On April 18, 1853 Stevens wrote to McClellan:

Among the other duties I wish to put into your hand is the construction of a military Road from Fort Wallah to Pugets Sound: An appropriation for it has been made of $20,000 and the Secretary [of War] has directed me to assign an Army Officer or Civil Engineer to the duty as my judgment might dictate.

I am most desirous that you should take charge of the road, as it is on the line of one of the Rail Road explorations, and I am sure instead of interfering with our general operations it will rather facilitate them.[8]

McClellan, then serving in Texas, traveled to Washington, D.C., where the two men spent a few weeks planning their westward explorations. McClellan was also tasked with exploring other passes in the Cascades with a view to their suitability for a railroad line.

Captain and later Major General George B. McClellan. Library of Congress, Washington, D.C. 3b50679.

Stevens left for St. Paul, Minnesota, on May 9. McClellan sailed from New York for the Isthmus of Panama on May 20 with explicit orders from Secretary of War Jefferson Davis, putting him under Governor Stevens' command. Davis also added:

It is important that this road should be opened in season for the fall immigration; you will, therefore, use every exertion to do so. Should it be found impossible to accomplish this, you will, at least, endeavor to fix the line of the road, especially through the Cascade mountains, and to perform such work on the most difficult portions as will enable the emigrants to render the route practicable by their own exertions, detaching a suitable person as guide and director to meet them at Walla Walla…"[9]

A Slow Start

McClellan arrived in San Francisco in mid-June and booked passage north. He reached Columbia Barracks[10] on June 27 and ostensibly began carrying out his orders. Showing "the slows" that exasperated President Lincoln a decade later when McClellan commanded the Union Army during the Civil War, the captain began to organize. He left the post on July 18, 1853 with a large caravan of sixty-eight men and 173 animals. The expedition covered two miles the first day out. That evening McClellan returned to the post and hired an additional packer. The next day was a bit better—the expedition traveled five miles. That night McClellan again returned to the post and hired six more packers. When later pressed to explain why he needed such a large contingent, McClellan answered that it was due to "the nature of what little information we possessed at the time in reference to the country we were about to traverse, the disposition of the Indians among whom we were to travel and other circumstances, which need not be mentioned."[11]

The rest of July and the first half of August were consumed in working his way east through the forestland south of Mount Adams to the Yakima River. The expedition moved at the rate of about five miles a day. On August 21 McClellan camped on the Wenas River, west of today's city of Yakima. Realizing the expedition needed more food, he sent Lieutenant Hodges to Fort Steilacoom on Puget Sound to obtain additional supplies. Hodges started the next day with a large contingent of men and horses. His route took him over Naches Pass and past Allen's road workers, where he stopped briefly to exchange news. A few days later, Hodges found himself with a number of broken-down horses twenty-five miles short of Fort Steilacoom. A letter was sent back over Naches Pass with a man named Lovelace informing

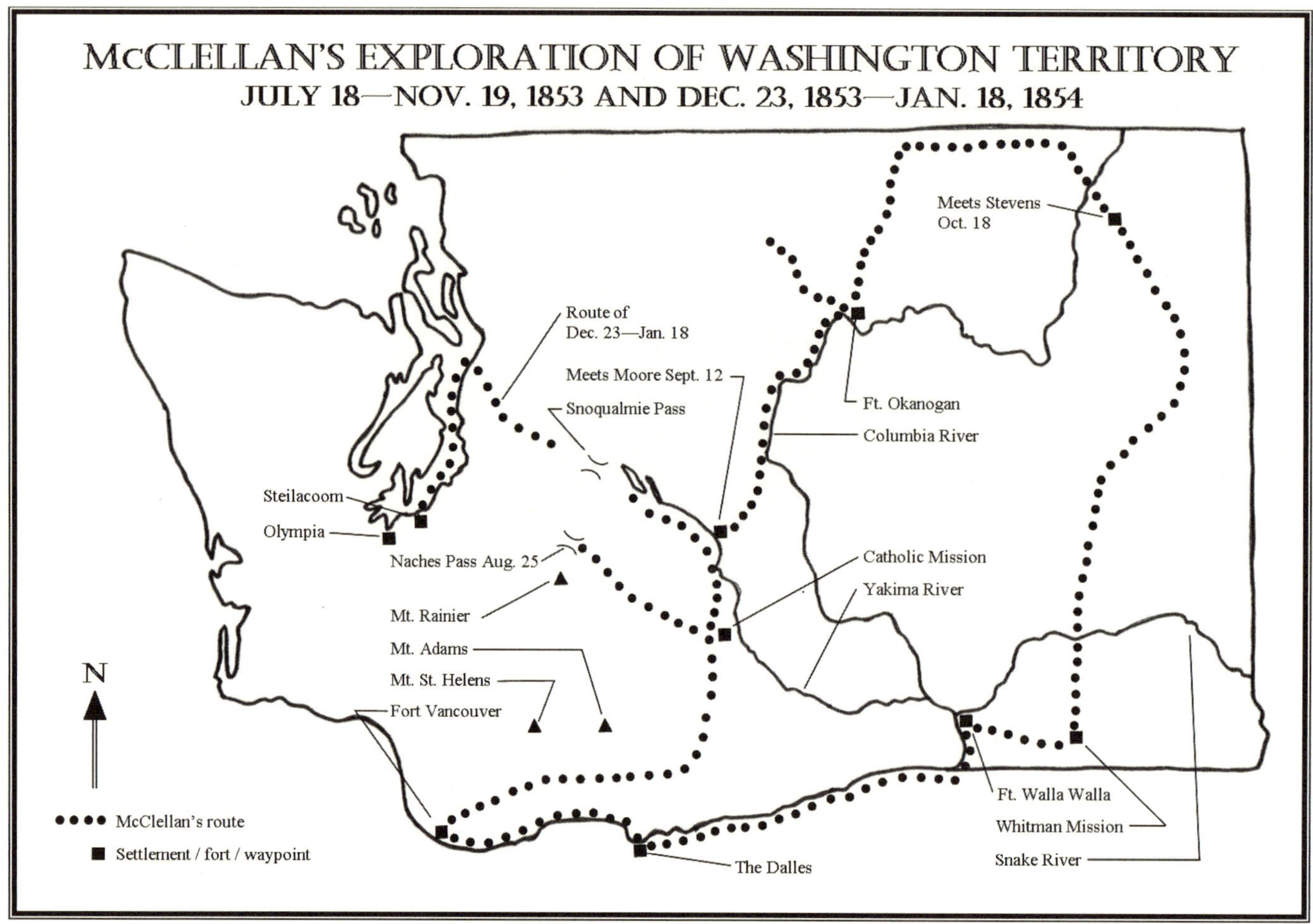

Map of McClellan's travels through Washington Territory. Map by authors.

McClellan of his situation and that the fort had only limited pack mules with which to replace Hodges' worn-out horses. McClellan received this note on August 31. Lovelace also told McClellan that Allen's workers had done a better job on the west side of the pass than was done by the workers on the east side.[12]

Finally, A Visit to Naches Pass

McClellan spent the days after Lieutenant Hodges' departure climbing Mt. Aix and exploring the nearby area. Finally, on August 23, he started toward Naches Pass. On August 25, accompanied by engineer J. F. Minter and six men, McClellan reached the summit, explored about a mile to the west and spent the night. The next morning McClellan started back to his camp on the Wenas, either forgetting or ignoring Jefferson Davis' orders about actually working on the wagon road. He arrived at Wenas late in the evening of August 29 where he conferred with Chief Owhi, brother of Kamiakin. On September 3 the party moved north to a base camp near present-day Ellensburg.

McClellan spent a few days exploring the headwaters of the Yakima River and returned to his camp on September 12. There he found Andrew W. Moore awaiting his return. McClellan had clearly disobeyed Davis' orders regarding the building of the military/emigrant road, and Moore's fortuitous arrival now offered him an opportunity to shed the responsibility entirely. He entered into negotiations with Moore and the next day signed a contract putting Allen's workers on the government payroll for all roadwork performed from that day on. Moore immediately returned to Olympia giving the road workers the good news as he passed them.

Lieutenant Hodges eventually reached Fort Steilacoom, and he returned to McClellan's Ellensburg camp on September 16 with the needed provisions. The return trip of over 150 miles was covered in just eleven days. Hodges and Moore would have passed each other en route, although neither mentioned it. On September 19 McClellan sent all but thirty-seven of his men back to The Dalles, Oregon, and continued north on a further exploration.

On October 18, McClellan and Governor Stevens reunited at Fort Colvile (about seventy miles north of present-day Spokane, on the Columbia River). As they worked their way south toward Walla Walla, the governor ordered McClellan to ride to Puget Sound over any one of the mountain passes that he had supposedly explored, to take barometric readings of its altitude, and to determine its suitability as a railroad route. McClellan did not want to do this and when the needed barometers failed to arrive he noted in his journal, "I therefore succeeded in having the project of crossing the mountains abandoned."[13] He erred, for on November 5 at Fort Walla Walla, Stevens urged McClellan to try the Naches Pass, promising him twenty fat horses and informing the captain that the emigrants got through with forty-eight wagons in September. McClellan again argued forcefully against the project, writing in his journal, "I objected so strongly to such a performance that I think it is now abandoned—for ever as I hope."[14] Stevens shortly thereafter left for The Dalles by canoe while McClellan and his men rode west along the Columbia River. Near present-day Boardman, Oregon, they learned that the late-arriving Mitchell wagon train had been forced to abandon its wagons near Naches Pass due to snow in the Cascades. McClellan noted in his journal, "so I was right & Stevens wrong."[15]

McClellan returned to Columbia Barracks on November 19 and remained there until December 12 when he left to report for duty with Stevens in Olympia. On November 24, the captain wrote his mother:

> We have to pass the winter at Olympia on Puget's Sound, a flourishing city of some 10 or 12 houses—fine prospect that. In addition I have to start again for the mountains as soon as we reach there—a trip of perhaps 3 weeks—in the rain & mud until we reach the Mts. & then snow...As I never saw a snowshoe in my life (except in a museum or a picture book) I don't anticipate much pleasure during the jaunt & am desirous of finishing it as soon as possible. As there are no houses in Olympia, that can be had, I expect to spend the winter in a tent—labored by the rain & mud—for you must know that we don't expect to see the sun anymore until next summer—except at rare and short intervals of time—it is raining almost constantly...I don't think much of it [the Pacific coast]—it is surely vastly overrated in every respect.[16]

When he arrived in Olympia, McClellan found, much to his displeasure, that the governor was still intent on surveying all possible routes through the Cascades. McClellan was ordered to explore the route from Steilacoom to Snoqualmie Pass, and he most reluctantly started for the mountains. On December 23 his party of six canoed up Puget Sound to Steilacoom, where they attempted unsuccessfully for several days to find horses and guides. With plans thus changed, McClellan continued paddling north up the sound to the mouth of the Snohomish River near the present-day city of Everett. They paddled up the river to within three-quarters of a mile of Snoqualmie Falls. On

January 9 the captain followed an Indian trail a few miles past the falls, encountered a foot of snow, and gave up the project. The dreaded snowshoes never came out of the packs, and by January 18 McClellan was back in Steilacoom. He spent the remainder of January and February 1854 living in Allen's Olympia cabin, writing and receiving reports.

Governor Stevens Fires McClellan

In a March 6, 1854 letter, an unhappy Governor Stevens ordered McClellan to turn over all government property and monies for the Cascade Road to Lieutenant Arnold and to "proceed to Washington City & report to the Secretary of War for farther orders."[17] McClellan finally received an order he acted on and wasted no time exiting the Territory. Olympia's *Pioneer and Democrat* reported that "Capt. McClellan of the Northern Pacific railroad exploring expedition left on [March 7] for Washington City."[18] Ten days later he was in Portland aboard the steamship *Peytona* bound for San Francisco.[19]

Meanwhile, Allen wrote to his family in Pittsburgh updating them on the governor's arrival and the progress he (Allen) was making in his quest for reimbursement for roadwork done before the government contract took effect.

Isaac Stevens, first territorial governor of Washington, 1853 to 1857. State Library Photograph Collection, 1851-1990, Washington State Archives, Digital Archives, http://www.digital archives.wa.gov.

> When I left off I was still on my claim, making out the cost of that part of the work done on the road, the payment for which had been guaranteed on the part of Government by Capt. M'Clellan, who, with the Governor, were duly expected to arrive, and to both of whom I was impatient to pay my "devoir;" not only to receive from the former the gross amount due, and to be able to pay the proportion due each of the faithful fellows who had so hardly earned it, but also to secure their united influence for the payment by Government for the *previous* labor done on the road; and, in addition, if in the Governor's judgment this road should be selected as a proper one for the 'military road,' I wished also (having had some experience in road making) to procure the contract for finishing it as such.
>
> The whole territory 'and the people thereof' were very anxious he should arrive, as much depends upon the Governor, and they were all anxious to know "what manner of man" he was. His antecedents were good, evincing a thorough-going, energetic, and pushing man, which impression his rapid but thorough survey of the northern route (of which we heard at intervals) increased.
>
> "Satis est." He came, and created a favorable impression on the instant. He is a small man, barely the average height, slender, spare, (which might partly be attributed to the hardships and severe labors coming overland,) but he was well knit—a world of endurance in his look, resolution and will. He bore about him the impress of a "man of mark." He is a very eloquent man, and made a brilliant speech in relation to the superiority of the Northern Route over all others, and the glorious future destiny of Washington Territory.[20]

Allen Meets with the Governor

Governor Stevens had left Columbia Barracks on November 19 or 20, continuing his journey north on the Cowlitz Trail to Olympia where he assumed office on November 25. Not all first impressions of him were as favorable as Allen's. Attempting to check into the Washington Hotel that had been reserved for the new governor and his entourage, Stevens appeared so travel-worn that he was told to wait outside until after the governor arrived. Presumably the error was corrected post-haste![21]

After a week or two had elapsed, I called on him and laid before him my propositions. He seemed much interested in the road, and admired the public spirit which (voluntarily, and relying on the good faith of the government to repay—or, if disappointed, no matter,) had determined, first to find, and then undertake such a stupendous work, as he flatteringly seemed to consider it. He requested me to draw up a succinct account of its inception, prime movers and actors, cost, &c., and he would embody it in his report to Congress. This (being an actor in it myself) I excused myself from doing, and referred him to others, and to the accounts published in our paper.[22]

Concerning the back payment for work done previous to the contract, he could not say, as that was M'Clellan's exclusive business; but the contract already made had been forwarded to him, and as far as it was concerned he was ready to pay, and accordingly did, by giving me a draft on Corcoran & [R]iggs[23] for $1,309.53, which I had to send to San Francisco to get cashed, and the payment of which to the men will occupy some time, as the poor fellows are scattered about considerably.

The total expense of making the road was $7,000, of which only the above $1,309.53 has been paid. The greatest difficulty seemed to be that Capt. M'Clellan did not know whether this would be the road ultimately adopted; as there was a pass, the "Snoqualmie," still further north, which he thought might be preferable; but as yet only the eastern part of it had been explored.

I endeavored to show him that, in consideration that this had been found a passable route by the passage of the emigration over it which had unanimously attested to it—that $7,000 had already been expended upon it—that over two hundred citizens were pecuniarily interested in it—that it entered the Sound at a central point—that it opened out on the open prairie, where the emigrant had good roads and the choice of going in any direction—in making up his judgment, these facts should have weight. Furthermore, in my conversation with the Governor to add cogency to the foregoing arguments, I took the liberty to represent to him that this road was a pet project of the people, who had called it the "People's Road;" that it had diverted the emigration from Oregon, and given an impetus to the progress and prosperity of Washington Territory, which no other object or project had; that nothing he could do at the introduction of his official career would be as universally acceptable or so commended by the people, or make him so popular, as his interesting himself in this matter.

Captain M'Clellan (prefacing his request by complimenting me on the clear and honest manner in which my statement for the $1,309 portion of the work he had assumed was made) asked me to make out an equally lucid and condensed account of the whole amount expended, and send it to Washington city, and he would recommend its payment by the Secretary of War. Governor Stevens also gave me assurance of his influence in the matter.[24]

Olympia Newspapers Weigh In

The *Washington Pioneer* jumped into the discussion with a rather long editorial on the People's Road. The editor gave the history of the road viewers, the raising of subscriptions, and the building of the wagon road, pointing out that $5,700 was still due the road builders.

As a matter of statistical information we would observe, that Mr. E. J. Allen has furnished us with data, from which we arrive at the following facts and conclusions: That the whole Territory has felt a deep interest in the completion of the road in time for the immigration of the present year, none will deny, and that our Governor, the United States Congress, and heads of Departments were equally solicitous for its completion, all thoroughly understanding the facts in the case, will readily admit. —That on the good faith of future provisions, there have been about fifty persons induced to become laborers thereon, on the guarantee of satisfactory adjustment of about 150 persons—interested by private subscription, and at a cost of labor, provisions, &c., of about $7,000. That on the last section of the road Capt. McClellan guaranteed payment to the chief of the party working on the western slope, the ordinary wages of the country, which upon investigation has been ascertained to amount to about $1300, and will recommend to the Secretary of War, that the deficit (some $5,700) be assumed for payment by the Government of the United States; which recommendation, we have every reason to believe, under the circumstances, as a matter of Territorial benefit and almost absolute necessity, will meet with not only the approbation of a majority of the federal and legislative officers, but of our entire community.... Fifty persons are justly entitled to $150 after the payment of the $1300 assumed as per contract of Capt. McClellan, and the question now arises, where is the balance (5,700) to come from? We will resume this subject again soon.[25]

On April 8, 1854 a notice appeared in the *Pioneer and Democrat:*

Those interested as laborers, or others, on the Cascade Immigrant road will find by reference to the advertising column, that E. J. ALLEN is prepared to make settlement of accounts for all work performed under the contract of Capt. Geo. B. McClellan, on presentation of the same.

In the same issue was this notice: "The workmen upon the western portion of the Cascade Emigrant Road are hereby notified that the amount due them upon the contract with Capt. George B. McClellan, can be had by presenting certificates of time to EDWARD J. ALLEN. Olympia, April 8, 1854." This advertisement ran once a month through January 6, 1855.

Allen did not give up easily. On February 14, 1860, the *Journal of the House of Representatives* stated on page 269, "The memorial of Edward Jay Allen and others, citizens of Washington Territory, asking payment of the expenses incurred in opening a portion of the military road from Fort Steilacoom to Fort Walla-Walla; which was referred to the Committee on Military Affairs." On December 13, 1860, the *House Journal* on page 73 again noted Allen's memorial.[26]

Despite lobbying by Allen and the editor of the *Pioneer and Democrat,* the U.S. government left no record of any payments beyond the initial $1,300 authorized in the contract signed on September 12, 1853, between Captain McClellan and Andrew W. Moore. Apparently, the federal government made no reimbursement of any kind for work done on the People's Road before this date. These expenses were borne entirely by the citizens of the territory.

Whaleboat Voyage

In December 1853, Allen and two friends sailed a whaleboat around Puget Sound—not an enterprise for the faint-hearted. Yet Allen, with a minimum of discomfort, accomplished his dream to explore the Sound, and encountered a virtual who's who of Northwest personalities in the process. The *Dispatch*'s typesetters obviously had difficulties with the various geographical names, especially those of Indian origin. The modern reader with map in hand will have little trouble following the adventures of a sea-borne Allen.

Allen's shipmates were Charles Weed and Charley Smith. Weed was born in 1826 in Connecticut, and was a member of the ill-fated *Georgiana* expedition to the Queen Charlotte Islands (see Chapter 2, note 95). He survived his capture, and settled on a land claim in Thurston County in 1853. In 1852 and 1853, he partnered with J. K. Hurd in the Olympia Bakery and Beef Market in Olympia. Weed served as a captain in the Indian uprising. He died January 10, 1896.[1] Smith remains an enigmatic figure in Northwest history. A different man, Charles H. Smith, had partnered with Michael T. Simmons in a mercantile business, and fled the territory circa 1851, after scamming Simmons out of several thousand dollars.[2]

Charles Weed, who was a participant in the *Georgiana* expedition to the Queen Charlotte islands in 1851, and accompanied Allen on the whaleboat voyage. Courtesy Washington State Historical Society, WSHS# C2009.0.1046

Allen's Claim," three miles below Olympia, Washington Territory, Dec. 27, 1853.
To Wm. H. Allen—*Dear Brother*:

"God's peace and love be with thee, wheresoe'er
This soft and spring-like air
Fans thy cheek or waves thy hair;
Whether its kiss comes to thee in crowded room,
Or out among the woodland bloom."[3]

As with me may it be with thee, "ever and alway!" Is not that a brotherly wish to come 4,000 miles? A letter freighted with a message like that ought certainly to be under special protection, and arrive safely—as I hope it will.

When last I wrote you I had returned from my four months and over arduous labors cutting the Emigrant Road through the Cascade Range, a brief account of which I sent you, and was busy on my claim, (whither I had returned to escape the embarrassing civilities and honors showered upon me,) rebuilding my house, getting out more timber, thereby gradually clearing the land; watching the tide of events; taking a rest, and alternating between my claim and Olympia—fast increasing her limits and extending her suburbs.

It had been my intention ever since I made my first home at Olympia to devote sufficient time for a complete exploration of the Sound with its network of inlets, bays, coves, ports, canals, harbors, and surrounding country—not only from curiosity, but with a view of permanent settlement at the most favorable and promising point; and though hitherto prevented by a succession of enterprises, which somehow seemed to follow on each completed previous undertaking as if by inevitable fate, and which had left me no one month's unoccupied time since I entered the Territory[.] I had never abandoned the intention, but longed for an opportunity to put it into execution.

A Welcome Opportunity

Very opportunely, I was afforded the privilege at this time—a chance I eagerly embraced. In my desultory active life I had chanced to come across Charles Weed, a high-minded, generous-hearted fellow, whom I was proud to consider my friend, who was going on the same expedition, and requested me to accompany him. The vessel was the whale-boat Knickerbocker,[4] of pretty good capacity and sailing qualities, commanded by Charley Smith, originally a Yankee sailor lad, who had strayed around here from Marblehead, whose life was of itself a romance, and of whom more anon. They landed at my place, which was all the notice I had—small time to deliberate. An Oregonian makes but little preparation—I simply gathered up my blanket and embarked.

There were but few hands about her—the "Skipper" and some sailors, Weed and myself—who occupied some such anomalous position as "ship's cousin," or "cook's mate."

It was a lovely morning when we started out. Sailing slowly past the left side of the inlet, till emerging into the broad channel of "the Sound" proper, our speed sensibly increased—

"Now freshening breezes swell the sail,
Low bears the vessel through the gale,
Bursts through the waves the pointed prow,
That loves the sunny foam to throw,
And thunders on before the wind
Long tracks of foam and whirl behind."[5]

Everything "auguring" a pleasant, safe and prosperous voyage—

"A bold, brave crew and an ocean blue,
And a ship that loves the blast,
And a go[o]d breeze piping merrily
By the tall and gallant mast."[6]

Let her proceed on her way, while I "discourse yees," on the past history and mysterious interest which has for three hundred and fifty years been associated with this "Land's End" of the "Olden Time."

I suppose that every one in his youth has formed in his own imagination, from studying geography or reading travels, his own idea of the "uttermost part of the earth," and has located it each and every one in different localities, ranging from zone to zone, and pole to pole. China and Japan seemed not many years ago to possess the requisites of mysterious vague distance, but in the "whirl of the world," they have become like unto other countries, no longer unknown. Thibet, and that religious, eccentric person, the "Grand Laam," (the Pope of that country) have become familiar to us. The oriental Cathay, Kamschatka, the cities of Trebizond and Timbuctoo, and that extensive blank space which not many years ago disfigured our maps, called the "Great American Desert," have all lost their shadowy interest, and have become tangible and real.

In consequence of much of my reading at a particular time of my "shining youth," consisting of voyages and travels, (having the good fortune, when the memory is most retentive, to come across that scarce work, "Hacklay's Voyages,"[7] I placed *my* "Ultimate Thule" at "Nootka Sound," and became much interested in the country and people of that far away place.

City born—brought up where "brick wall and heaven is done up in soot," there was something peculiarly cool and cosy in the manner and style of their habitations—a large room under ground, with a notched pole in the centre, forming thus means of descent and ascent. I have seen pictures describing their interior.—The beds were composed of dry moss and other warm materials, and ranged above one another in tiers, occupying one side of the room. It was a pleasant conceit to me then to fancy myself occupying an upper bunk, and after eating my fill, pelting the lower occupant with pig nuts.[8]

Read without a map, I never definitely defined its locality, and frankly confess was ignorant of it till a later period. None of the mysterious interest attached to it has been dissipated from the since ascertained knowledge of its being so near Puget Sound. I should very much enjoy a visit to my pig nut friends, and some day may go. We think nothing of such trips here...[9]

No description of country, however forcible and vivid, can give the reader a "realizing sense" of that described—without the additional accessary of a

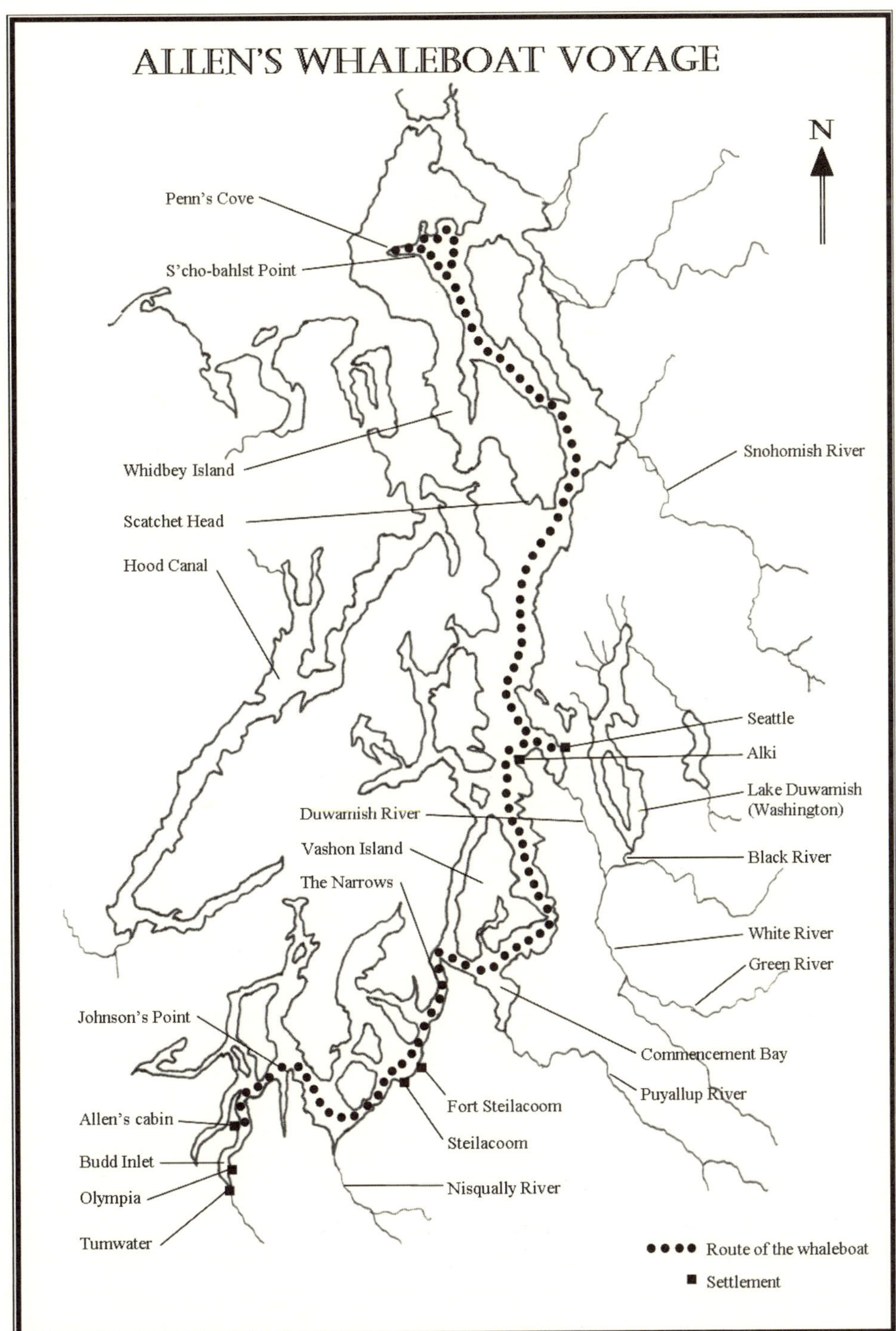

Route taken by Allen and friends in their whaleboat. Map by authors.

Sound and numerous indentations resembled "a damaged fig, the stem standing for the Strait"—a very good comparison! I send you, with this, a map of my own construction,[10] done of course not artistically, but from actual knowledge; the distances, latitude, and general outline being correct—a mere outline map, but upon which all claims and the names of the occupants, on Budd's Inlet, are noted, and all the towns, principal points, fisheries and sawmills, on the Sound, are marked, so far as I have explored.

Northward Bound

You must now, (in resuming the subject once more,) imagine that, while thus engaged in "discoorsin" about its past history, the good bark has not been stationary; but, once entered upon Puget Sound, passing Smith's Bay,[11] and keeping up the right hand shore, a "day has been consumed, a night passed away," and skirting past Johnstown's Point,[12] which is fifteen miles from Olympia, extending, like a peninsula, commands a lovely and extensive view up and down the Sound. I was enchanted with this place, and—would like to be its possessor. Fifteen miles further down, on the same side, we come to Nisqually, the place Wilkes, of all places, so admired for its beautiful scenery, and which has not changed since his time in that respect. The shores are very high, rising two hundred feet—the banks lined with a fringe of oaks, ash and pine, as described in my

map—and a map of Puget Sound is hard to procure, there being none extant to my knowledge, except Wilkes', which I have tried in vain to obtain for you. I laughed some at your response to my recommendation to consult the maps in the ordinary atlas, (which I ignorantly supposed were reasonably exact,) that the previous letter, with the prairie for a boundary, and Mt. Olympus, covered with snow, for a back ground.

It was from here we proceeded to view and afterwards cut the Cascade Road, and from the prairies back we obtained the view of three mountain peaks at once—Mt. Olympus,[13] Mt. St. Stevens,[14] and Mt. Rainer—which, rising in the East, towers to a height of 12,330 feet![15]

From Johnston's Point to Stilacoon, a distance of fifteen miles. Stilacoon is destined to be of future great importance, as it has an advantage over Nisqually, in possessing a good harbor at all times safe, beside being surrounded or rather backed by a fertile country, and being the terminus of the Emigrant Road, leading to Fort Wallah-wallah. There are two towns here, upper and lower Stilacoon.

Immediately below Stilacoon is the "Narrows,"[16] so called because at that point the whole waters of the Sound are compressed within a space a mile and a half, with a current like a mill race. These narrows forms the limit of Puget Sound proper, and below commences Admiralty Inlet, with its bays, islands and harbors—Puyallop being the first upon the Eastern side, around which we went a sailing, distance from Stilacoon twelve miles. This is the bay called by Wilkes "Commencement Bay;"[17] it is a very handsome sheet of water, as indeed all the bays are—and, in common with all, has many small streams emptying into it. The Puyallop River[18] empties in this bay, and can be made navigable by clearing out a few drift rafts, for forty miles, perhaps. The country along its borders affords rich lands, and upon the bay are facilities for large salmon fisheries; there is, however, a drawback—the receding of the tide leaves large mud flats, which extend in patches as far as the point near the Narrows.

My friends, Messrs. Swan[19] and Riley,[20] have here established the first regular salmon fishery upon the Sound, and are doing well, though they complain of their net being too small, only averaging, they say, ten or twelve bbls. each haul! the average weight of the fish being twelve lbs., though some go as high as thirty-six.—They are seeking a location for a cod fishery, which (when found,) they will engage in largely, being enterprising and industrious men, qualities which are, indeed, the general characteristics of our population. They have a beautiful location, commanding a most lively and extensive view down the Bay; looking toward the eastern side of Vashon's Island and extending almost to the point of Duwarrish[21] Bay. Sailing up, or rather going out of Puyallup Bay, we had a glorious glimpse of Mt. Rainer, while away to the Northward I saw for the first time Mt. Baker.[22] Vashon's Island[23] comes up to the North of Puyallup Bay, only a mile below the Narrows, and is about twelve miles long by one and a half miles wide, with high banks, plenty of water and heavy timber, but soil of an inferior quality, and some little prairie; we camped on it that evening, or rather "lay to." At the foot of Vashon's Island, and about three miles below comes in Duwarrish River, which receives a portion of the White river of our Emigrant Road, while the other has broken through and empties into Puyallop. Curious, is it not?

Alki and Seattle

The Bay has, at it[s] debouchure Admiralty Inlet,[24] thrown out a long tongue of land, upon which is located a town, just seven miles from the foot of the island[,] formerly called New York, but altered since to the more appropriate one of Alki,[25] taken from the Chinook language—meaning "hereafter," or "after a while," equivalent to the *poco tiempo* of the Spanish. The view from the point in extent and beauty, far exceeds that from the foot of Puyallup Bay—but unfortunately it posesses no other advantage, for the country immediately back of it is inferior, except that part which margins Duwarrish Bay, which will doubtless be diverted [t]o the town of Seattle,[26] situated on it and just five miles from Alki.

There is a small saw mill here, belonging to Mr. Terry,[27] of ordinary power—circular saw, cuts five hundred feet in three hours; he never has sold any lumber at less than twenty-five dollars per thousand feet, and oftener at forty dollars. We laid in some fresh provisions, for which Mr. Terry would accept nothing, and I could not refrain from repeating to myself, as I left his hospitable house, "there was a jolly miller, once, lived on the banks of Dee."

Seattle, as I have said, was four or five miles farther up, on the other side of the bay. The tide was out, and left a margin of white glittering sand bars, against which the waves rolled murmuringly. The day, tho in the month of December, was an unusually mild one,

Typical whaleboats, powered by oars and sails, on Puget Sound with Mt. Rainier in the background. *The West Shore*, May 1884.

even in this climate, (what a contrast to last winter!) the skies clear, the water still and quiet; when it was deeper, it was so pellucid that it seemed to reflect the surrounding objects like a mirror.

> "Against its ground
> Of silvery surface, rich hill and tree,
> Still as a picture, clear and free,
> With waving outline, marked the coast for miles around."[28]

Disembarking from our gallant craft, a little above Alki, we directed them to proceed and wait for us, while we walked up the gravelly beach to look at an Indian encampment; our curiosity gratified, we came opposite Seattle, and waited impatiently the arrival of our craft; we had concluded to camp out, on the smooth bar, when (at midnight) it "hove to," too late to go over, and so, in preference to sleeping on board, we spread our blankets on the beach,[29] and were soothed to sleep by the deep murmurs of the waves. We passed a pleasant night, and arose refreshed in the morning, tho it rained heavily—which may appear strange to you, but no less true—so indifferent were we to it that, altho we had a tent, we cared not to use it; our blankets were water proof, and the night was dry.

Next morning we crossed over to Seattle. It is rather a pretty place, situated pleasantly for a town, but they have built it (instead of fronting on the bay,) to face a street about a block from it, which brings the rear of the building is in view first. Their idea I believe is, however, to grade the bank, and then build fronting the bay. It contains about twenty-five houses, and is increasing rapidly. Alki has but five or six. I do not exactly like the locality as well as some others I have seen.[30] There is a mill here, somewhat larger than the one at Alki, running with a circular saw, which limits their business as they cannot cut large timber. Two vessels were loading with lumber, while we were there—a pretty sight, lying close to shore.

Farther up the bay the Durramish River enters, and is said to be fed from a lake [Lake Washington] only five or six miles from Seattle and immediately back of it, taking a circuit of twenty five miles before entering the bay. The lake is said to be full of fish, and forty miles long. I know not if this is true. Wilkes' survey was made in May and he describes the surrounding country as being "covered with a fine growth of trees, with here and there a prairie interspersed, with its honeysuckels and roses just in bloom, resembling a well kept lawn; the soil is capable of any kind of production, the woods alive with game," &c.

Whidbey Island

Leaving Seattle and passing down the bay, we emerged again into the inlet, and steered for Whitley's [Whidbey] Island, which loomed out faintly in the dull light, Skaget's Head,[31] with its grey cliffs, standing out in bold relief against the dark background of the green shores.

We disembarked and camped on the mainland—as usual, on the sandy beach—enveloped in our blankets, prone on our backs, with faces upturned to the calm, quiet heavens.

"Let him who crawls, enamored of decay,
Cling to his couch and sicken years away—
Heave the thick breath, and shake his palsied head,
Ours the fresh turf, not the feverish bed."[32]

From Seattle to Skaget's Head is 30 miles.—It is at the extreme Southern point of Whitly's Island, and is a cliff or cliffs rising boldly from the water, forming a prominent landmark as you approach it. In the centre is an indentation, or cove, forming an excellent harbor.—There are one or two salmon fisheries established here, but not as extensive as Swan & Berry's establishment at Alki;[33] they employ Indians to do most of the labor, (filling the barrels, &c.,) who procure the salmon at the Falls of the Swohomish[34] river, which empties some few miles below, bringing them to Skaget in canoes.

The timber on the Island is poor and here is reversed the ordinary rule of judging good land; generally the prairies are the poorest, and the wooded lands the best; but here it is *vice versa*.

Formerly a powerful tribe of Indians lived here, but all have faded away—their old chief sleeping his dreamless sleep on the brow of the cliff.

"Tranquil now he sleeps at last,
On the broad cliff's bosom cast."[35]

The traces of their habitations have faded from earth; unlike any other unfortunate race, swept away in the march of nations, no relics are left—ruins, religion or language—

But the doomed Indian leaves no trace;
No lofty pile, no glowing page
Shall bind him to a future age,
Or give him, with the past, a rank.
His heraldry is but a broken bow,
His history but a tale of wrong and wo—
His very name a blank.[36]

The settlers here have done one commendable thing; instead of naming the embryo towns after the manner of the States, where two Syracuses are very often within speaking distance of each other, and perhaps a Rome or Athens will lay over against them—or as they do in California, (and significantly enough, too,) name their places "Murderer's Gulch," "Bloody Hollow," "Squatter's Point," "Belcher's Ranch," &c.,—time, as you see, returned the original names, which are always euphonious and expressive. I am in hopes, too, that when our Legislature meets they will carry out the good work already begun, and instead of the mistaken policy of making *pariahs* of the Indians, extend to them the benefit of our protection and laws. There are but few left, many of the earliest settlers, and now wealthiest and most useful and respected citizens, have years ago married among them. The tribes about the Sound, Hood's Canal[37] and Admiralty Inlet, are quiet, peaceable, inoffensive—but degraded; will steal, (as I know to my cost,) but are not turbulent, as the warlike Indians residing further north, in the British possessions.

I must not forget to mention that both at Skaget's head and Alki, we received from the proprietors (and especially from Dr. Maynard)[38] of the latter place, that courtesy and hospitality so characteristic of the people here.

We passed down by the mouth of Snohomish river, and had time permitted, I would have ascended, and admired the magnificent falls[39] upon it. It is said to be the largest river emptying into the Sound, and falls a distance of 252 feet. It must be a glorious sight. You see we have all varieties of scenery here—distant ocean, fresh and salt water lakes, coves, ports, bays, inlets, natural ship canals, woods, prairies, hills, valleys, plains, cliffs, headlands,—and we can pitch you in the crater of some volcanoes, happily all extinct but one.

Penn's Cove

We sailed on, and camped in Penn's cove,[40] thirty-five miles from Skaget's Head. This cove is a beautiful harbor, nearly cutting the Island in two, and at the head of it is a prairie, and a little town called Coveland.[41] It is rather a pretty location, but I do not admire the Island as many do. I do not much fancy living upon an island at all, even though it is as large as Whitly's. The best land lies about the cove, but it is nearly all used up now. Wilkes, who saw it in May, in the flush of summer, was enraptured with it. It was then inhabited by Indians; there was a Catholic mission[42] established among them, and much land was under cultivation; potatoes and beans were growing—wildflowers in abundance, and the whole surface covered with wild strawberry vines, the fruit large and of fine flavor. The island is curiously shaped; in going down it seems to close in the shore beyond—being very winding, not varying much in width but following all the curvatures of the shore. This singular conformation of the island creates a curious tidal arrangement. When the tide is coming *in* the current flows outward, and when it is running *out*, the current sets inward; this is upon the eastern side; upon the western it sets as usual, thus flowing around in an extensive eddy, as the island is nearly fifty miles long.

The scenery, both on the island and the main shore is very beautiful, and on the other side equally, if not more so, altho I only know of it from reports of those who have seen it.—Wilkes compares is [it] to the scenery of the Hudson river, particularly about Poughkeepsie, and above that place—the distant highlands reminding him of the Catskills.

The singular effect of the tide, the striking features of the scenery, changing with every curve; the beauty of the day; the calm grandeur of the blue sky; the quiet pervading—all conspired to cheat the fancy into the belief that the "changing panorama" was unreal.

"Earth, water and air, and the glorious sky,
With a holier calm were impressed—
And each island that lay in that glittering bay
Seemed a Paradise of Rest."[43]

This part of our voyage was perhaps the most pleasant of all, reminding me somewhat of our passage down Lewis' river, tho we were now in a rather larger craft. It was pleasant, with sails idly flapping the masts, and just enough headway to give her steerage, to take advantage of the singular current, and with no effort thus go sailing by. It would be impossible to describe the various kalideoscopic changes we glided past.

Altho so easy getting *to* Penn's Cove, it was not so easy getting *away*—nor did we regret it much; we remained three days there, the bark, anchored in most "inglorious inactivity," waiting our good pleasure and convenience. We pitched our tent on the gravelly beach, and "did whatsoever seemed good in our sight;" after the first day I did not ramble over the island any more, but spent my time near the tent, watching the curious effect of the tides, receding and advancing like a coy lover—the white sails of the passing vessels—or, glancing heavenward "the cloud armada" sailing down the sky; while in my reverie, came thronging a thousand thoughts—some cheering, exhilerating, hopeful and inspiring, and (typical of life with its lights and shadows,) some were sorrowful and sad—for I am not the least ashamed to confess that even with an apparently buoyant temperament I cherish a gentle melancholy, to keep it within limits, and ignore no thoughts, be they sombre or "lichtsome" that come unbidden in my "lonely musing." There is that, in every man's experience, that in its gayest words, if suddenly recalled, must "give him pause"—and thus, alone on that solitary beach, the vision of a home four thousand miles away, and the thoughts of one who went out into the great world and "never repassed that threshold more," came upon me with a subduing influence.

Upon the third night passed upon the island, a wind sprung up, and (striking our tent with as great celerity as an Arab,) we embarked once more, set sail, and ran out of the cove—and *such* sailing I never saw before. My nautical experience having been hitherto confined to paddling a skiff round the riffles in the home rivers, and an after trial as captain of a ferryboat at Fort Boise, I was totally unprepared for this new mode of gliding thru the waves at an angle of forty-five or fifty degrees—now on one side, and now on the other—shifting our position alternately, to keep up an equilibrium. The wind was blowing due west, creating a heavy sea, and we had only achieved three miles, when nearly capsizing her in the attempt, we brought to at a little sheltering cove. Tenting on the beach as usual, we found the wind had changed in the night, and we were compelled ingloriously to "beat out" of our shelter. We got along very well, until (in rounding past one of the numerous headlands,) the wind took us, and drove us down the bay, and we were fain to take in sails, and betake us to our oars. The wind increasing made my stomach rather qualmish—and, for a time, there was a simultaneous "heave" *on* the long sweep, and *from* the aforesaid stomach. We reached a harbor by "main strength and awkwardness," and this gave me an opportunity to examine the old ruins of the immense houses upon Schroreckbin's Point,[44] which in Wilkes' time, only eight years ago, was peopled by a civilized race of Indians, who cultivated the soil, but who have now almost vanished from earth.

An Abandoned Indian Village

Their houses were built somewhat after the manner of the Queen Charlotte Indians. There were ruins of five or six, one of which we examined minutely. It was well framed with heavy squared timber, and boarded up with large boards, split from cedar. It was about six or seven feet high, but went down a foot or so under ground, making it about ten feet; the roof, slanting up to the center, to the height of twenty feet, made a capacious room. The center beam was a large tree, two feet in diameter and one hundred and fifty feet long. How they contrived to place such a heavy piece of timber in its position, is a marvel. They all lived in it, and it was considered impregnable to assaults from the neighboring tribes. Its width was eighty feet.

It is silent and solitary now, deserted by the Indians, who have a superstitious feeling regarding it, as a great many of them, with their chief, died here—carried off by that scourge of the red man, the small-pox; the others moved away—and the winds moan mournfully over the ruins.

Their grave-yard is near by—the bodies being enclosed in wooden boxes, set on end—almost incorruptible in this pure atmosphere.—Some of the head-boards, carved fantastically, which have an exceeding picturesque appearance as seen from the bay. The space around is strewn with all an Indian's wealth and treasure—blankets caps, carved pipes, shells, streamers of crimson, etc, etc.

The man[45] occupying the claim below is about using the old house for a stable for his horses—and soon the head-boards will be converted into cattle posts. His claim is a beautiful one, sloping gently down to the beach, and merging into a prarie back; he has left some yew trees standing (they grow to a great size here,) and a cluster of ash around his house; one or two deer hanging from the branches, giving it quite a frontier look, which was mellowed down into more of an aspect of civilization by the flitting around of a *dear* little wife; who, engaged in household affairs, would (in passing,) give him such deferential and loving looks as were pleasing to witness, yet absolutely "aggravatin'."

Southward Again

Leaving this, we floated out from the beach, and passing under the bluff, took a last look at the graves of

Saratoga Passage between Whidbey Island and Camano Island. Allen and friends likely camped on a beach similar to this the first night after their escape from Penn Cove. Photograph by Dennis Larsen.

the extinct tribe. Our voyage, that day, was anything but pleasant. It rained part of the time, making of us uncomfortable moist bodies. The rain ceased and a breeze sprung up; provisions were short, we live on dry salmon; the breezes increases, carries away a sail. We battle with head winds—manage to get ashore somehow, without having to "hoist signals of distress." Once more, the "well secured," and "'neath a headland safely moored," we made a fire, dried our blankets, had a supper, and were rather comfortable. The ground is yet a little dampish, and the hoarse murmur of the winds betoken a storm. We crowd a little closer together. The moan of the winds mingles with the roar of the surf on the beach, and both talk a language intelligible to each other, I doubt not, as they keep up the converse till morning—which, unexpectedly, bursts upon us all goldenly, a bright beautiful day, such as *you* don't have till May—and, as we had rather overslept ourselves, the sun was high up in the heavens ere we rose.

"Blest power of sunshine, genial day,
What life, what power is in thy ray."[46]

To our surprise we find ourselves but a little distance from Skaget's Head. The fishermen receive us most heartily, have plenty of provinder, consisting of salmon; we diminish the "pile," somewhat—and try to make a start, but the wind is again against us, and we resolve to wait for the tide. I pick up the "Life of Wickliffe," conceiving therefrom a higher respect for the fisherman who owned it, got very much interested, and wonder, had he lived till now, if his ire would not have arisen at the attempt to ostracise from our schools the Book containing the doctrines he was willing to give up his life for?

The wind lulls again. The tide *has* turned. We start out again; the wind rises, soon as we get fairly out; go for four miles like a streak, all hands leaning over the high sides to keep from capsizing—another sail gone—we right up tremenduous "swell"—Neptune and Boreas, hitherto good friends, running opposition to each other.

Reached Alki by noon; renewed our stock of provisions, for which no pay would be taken, as usual; had a good run till toward evening—when, as we were somewhat crippled in our rigging, we "came to" again, and camped on the beach. By the time we had built a fire, pitched our tent and eaten our supper, it was too dark to observe at what kind of a place we had landed.

We passed a pleasant night, the expected storm having passed to other regions. Before we closed our eyes, we saw the sun had set over the distant range of mountains—

"And, by the bright trick of his golden oar,
Gave goodly promise of a glorious morrow"[47]—

a promise fully realized. How I wished you be-smoked and be-fogged unfortunates could enjoy it with me. Think of it! December—the annoying time with you—"when noses bloom purple in the blast"—and so like unto Spring here. The day was not only exhilerating, making "our hearts glad within us," but on looking around, we found we had camped under a bluff, jutting out in the bay, perpendicular on one side—with mosses and wild-flowers, and creeping vines, checkering its surface. There was a cove found in the same side, where the "still waters ran deep;" the ascent on the other was gradual to the top, and grassed to the margin. There was a large grove on the apex, of trees, the like of which *you* do not see, and wishing to get the view, we ascended and were soon under its solemn shade.

"Here arches high the forest's golden ceiling,
And licks the heaven of blue—
Save where a dim and holy light is stealing
The twining branches thro;
Here, mossed with age, stands many a heavy
column
To prop the mighty hall—
Nought breaks the silence, undisturbed and solemn,
Save when the dead leaves fall."[48]

Our "skipper" here gives us a snatch of his life, which has been an eventful one, and of wandering too; he had never told his life connectedly—only in scraps, and in a humorous way, as if it was all a joke, though I could not appreciate the humor of some of it. Being after many adventures, "a little short," he shipped before the mast, to come around here—unfortunate selection of a vessel, rather "hard lot" for a crew, mate a tyrant—a belaying pin was his weakness, his peculiarity being to bring it into close contact with a softer substance, and always selecting for his experiment the weakest and youngest—and our skipper not being twenty yet—he was the victim generally. A blow aimed at his head misses it and breaks his leg; there is no doctor on board, and no one else can set it. He lays in his bunk, setting it himself. Comes around here, buys this little whale boat to trade up and down the sound, to carry lumber, and *go on exploring expeditions*!

The skipper's story being told, we took one lingering last view of the panorama spread out before us, and unwilling to lose the benefit of the tide, descended to our craft musingly. The beach reached, we strike tent, heave anchor, hoist sail, and are afloat again. We met seven vessels in the Narrows, all in full sail—glorious sight. Hailed at Wilton's[49] claim to heave to, and dine with him, but declined reluctantly, and hurried on to Stillacoon. Called to see my friend Captain Balch, the proprietor of the town, who contributed so liberally to the executio[n] onf [of] the emigrant road. I met here quite unexpectedly, another old acquaintance. Do you recollect the family who suffered so much coming over the plains, and who were so kindly treated by the different trains? You may recall them to your memory by the incident of the affectionate-hearted little wife grieving over the necessity of leaving behind a favorite article of domestic use. We were invited to supper, and almost overwhelmed with kindness. A woman never cries in vain. There was "no telling" how many associations of past household joys and sorrows were linked with that skillet—*it was not left behind*;[50] surreptitiously concealed, it accompanied them all the way to their new home, and was now triumphantly waved before me, with the accompaniment of a hot cake and a bright smile, puzzling me to tell which was most appreciated! There are two places here, Upper and Lower Stillacoon, both well situated, the Lower rather more prosperous.

Fifteen miles brought us to Johnston's point, the beautiful place I have described before—the most beautiful site, for a residence merely, on the whole Sound. We landed and called on the Doctor, who owns it, who was going to the States and was willing to sell out; asked $2,000 for it, and wished I could have afforded to buy it. Talked so much about it to Charles Weed, that I have persuaded him to secure it.[51]

Returning homeward, we encountered a sight—a small steamer significantly named the "Fairy,"[52] come to run as a packet—too small, however, to do any business; not much larger than two ship's yawls put together, carrying only three tons—but the advent of even this little steam vessel created great excitement at the Sound.

There is a great opening here for steamboat operations. The emigration w[i]ll push toward the Sound; which, before the back country is taken, will be all settled along its banks. With your ten years experience, as one of those "who go down" the rivers in steamboats, and do business "on the great waters," I should think you would make a fortune. There are plenty of moneyed men here, who would engage in the undertaking with you.

A few more hours brought us in sight of my home, and sailing past it we proceeded to Olympia, which I found much improved, and everything looking bustling and prosperous—and to extend this voyage around Puget's Sound—a misnomer, by the by, as we did not sail *all* around the Sound, but to make good our title to this letter, I extended it, as you see, to Admiralty Inlet.

A Description of Puget Sound

Strictly speaking, Puget Sound is only that portion *laying South of the Narrows*. Below are Admiralty Inlet and the Straits of Fuca. The Straits of Fuca I have never seen—nor that portion of Admiralty Inlet beyond Penn's Cove, from which place to Bellingham's Bay is a distance of forty miles. That bay possesses a fine harbor, rich land behind it containing coal, which is now being worked by the Marmosa Company.[53] In Puget Sound proper there are two Inlets or Bays, supposed to be as large as Budd's Inlet—called South Bay, which is to the East, and Ells' Inlet, on the left of Budd's. (My claim fronts on Budd's and runs back to Ells' Inlet.) These have never yet, I believe, been explored; both of them contain oysters. At the foot of Ells' Inlet, Oyster and Skukum Bays, lying across their mouths, is a large Island,[54] which I mention simply because it is so large; and nothing known concerning it—for these seas are full of islands, that empurpled beautiful waters.

Hood's canal opens near Port Townshend, which is a fine harbor, three miles and a quarter in length and a hundred and thirty-four wide, free from danger, well protected from storms, the soil around it considered good and productive. The bay is upon the left hand side, and near the mouth of the strait. The canal stretches, in a kind of circular triangular shape—bending outward, leaving a large extent of country between it and Admiralty Inlet and Puget Sound—for it is as long as both put together—till it approaches and leads within two miles of Skukum Bay, or more properly Hammersly Inlet, which is a portion of the waters of Puget Sound. Over this short space the Indians have a portage, where they carry provisions. Some day, may-hap, the two waters may be united by a canal. A large stream empties into the bend of the Canal, by which communication can be had with the Chickalus [Chehalis] and Columbia River's and valleys, and Gray's Harbor on the Pacific Coast, not far above Astoria, if it should prove to be a good harbor, which I very much doubt will ever be the case.

At the point of land, between Skukum and Oyster Bay, will (in my humble opinion) at no distant day be an immense place—which, with the map before you, I think will be apparent to you. It is within three miles of the head of navigation, by Hood's Canal, and commands all the Chickalus Valley, over which I crossed in crossing to Puget Sound.

The bank along the Canal average one hundred feet high, formed of stratified clay, according to Wilkes—light gravelly soil covered with a growth of pines. Toward the North, the shores grow more bold and rocky, with a fertile soil, formed by the alluvial deposits from the Mt. Olympus Range. Toward the head, the shores slope to lowland, with a good soil.—The water in the channel is too deep for anchorage, but there are many good harbors in it, with small streams emptying in them with sufficient power to run mills.

A few more remarks about the safety of the entrance to all the network of Inland Seas.—Vide Wilkes, again:

"The Straits of Juan de Fuca may be safely navigated. The wind will, for the greater part of the year, be found to blow directly through them and generally outward. This wind is, at times, very violent. The shores are very bold, and anchorage to be found in few places; we could not obtain bottom in some places with a sixty fathom line even within a boat's length of shore."

Speaking of the Inland Seas he says:

"The shores of all these inlets and bays are remarkably bold—so much so that, in many places, a ship's sidds [sides] would strike the shore, before the keel would touch the ground."

"Nothing can exceed the beauty of these waters, and their safety. Not a shoal exists within the straits of Juan De Fuca, Admiralty Inlet, Puget's Sound, or Hood's Canal—that can in any way interrupt their navigation by a seventy-four gun ship. I venture nothing in saying there is no country in the world that possesses waters equal to these."

Is it not one of the most extraordinary things in the world that the general reader, the great mass of the people, and (from the indifference with which our government gave up a part of it to England,) our public men, also, should be so long ignorant of the importance and magnitude of these wonderful waters, and this beautiful country!

It is partly attributable to the jealousy of our neighbors, in lower Oregon, the supineness of our own government, in the settlement of the boundary question, and (I am sorry to be compelled to say,) the incorrect and prejudiced accounts of some of our missionaries; who, with commendable prudence, *entered themselves each* 1,000 *acres of land*, in view of its eventual progress.

The most potential, of all the influence that has conduced to this result, is the ignoring by the government of all Northern measures, and the strenuous opposition of the "chivalry" and "slavery propagandists," to all and every measure that would tend to disturb the strength of their "peculiar" (thank God, it *is* peculiar to them!) "institution," aided by Northern "*doe*-faces"—the *true* word, as shrieked at them by John Randolph! The South can have any amount of appropriations to find an *impassible* pass over the Sierra Nevadas, where a party caught once in the winter, camped on the snows—on which, as they gradually melted, the square frame-work of twenty-five feet in length of logs, they had built their fires upon, gradually sunk, till it rested on the level ground—a square room, of twenty-five to thirty feet, by forty feet depth, up the sides of which they cut steps in the snows to ascend! Or over the "howling deserts" of Sonora, fit residence for the gaunt wolves, tarantulas, and centipedes, which, (as we used to say at school, when reciting our geography lesson,) are its "chief productions."

We have Fremont's Expedition, Stansburry's, Emory's, Gumman's, Beal's—*cum inultis alias*[55]—but none (till Stevens',) for the North. For the purpose of blinding the eyes of the public to the immense cost of these surveys, and the partial distribution of the appropriations, it has been represented that some of them are mere private undertakings, costing but little, and others conducted on so economical a scale that the entire outfit consisted of but a rifle, tin-cup, extra leather shirt, a cracker, chunk of cheese, and a "sassenger" [sausage]—which is all "bosh," as the appropriation bills will show.

I will not be very much astonished if the report of Stevens, and his effiecient assistants, will not astonish the whole country. It is said he has already found *a pass* 1,000 *ft. lower than the South Pass*, a little further North. We know what a country it is, from the Missourium there. The most, and only difficulty, is a passage over the Cascades—which affords two, ascertained to be possible in December. The Californians, so soon as they see the practicability of *this* route demonstrated, will be anxious to unite fortunes in concentrating on the Central Route and in assisting the Oregonians in bringing a branch up the Rogue river, Willamette, Chickalus valleys, to Puget Sound, which is *nearer to Japan than any other latitude on the coast by* 1,500 *miles!*

The emigration to Washington Territory will be very great, as soon as its advantages and facilities for commerce are known. Already are capitalists and merchants swarming in.—Those who wish a home on the *shores* of the Sound will have to come soon, as all who enter the Territory tend thitherward, and the donation law will soon expire by limitation—unless renewed by our Legislature, which I hope will be done.

You recollect in one of my letters I prophecied that gold would be found in the Cascade Range—and I did it with as much faith—though not predicated in the same geological knowledge—as Sir Roger Murchism, the celebrated British Geologist, did before an audience the discovery in Australia. It has been since verified. It *has* been found in the Cascade Range, near Mt. Rainer, and by Captain M'Clellan, one of Governor Stevens aids upon the Yemass [Wenas] River, which empties into the Yakino [Yakima].

This will cause, no doubt, a great rush to the country, and I am sorry for it—as it will bring the kind of population we don't want, and I fear will produce the lawlessness and dreadful scenes enacted in California, the inevitable effect of a severance of all home ties and influences, consequent upon

"The gold and silver's dreary clanking
On the metalic heart of man."[56]

I returned from my trip, more fully impressed with the future prosperity of Washington Territory, and irate and indignant at the treachery to the North, apparent in surrendering all we claimed, when treating with England. Had it been a cactus desert in Mexico or Cuba, we would not have so ignominuously "flunked." I do not believe, really, that if England were to offer to sell it, for half its value, the present administration would buy it, not even Vancouver's Island.[57]

Allen for the Senate

This week Father took in a bill to James Tanner [a local merchant], he spoke of you, saying you would be a... Senator. —Rebecca Allen to Edward Jay Allen, September 27, 1853

The act creating Washington Territory directed Governor Stevens to call for elections for the various offices and positions in the government of the new territory, to be held at a time of his choosing and directed by his appointees. The governor was empowered to set the number of members of the Council (Senate) and House of Representatives to which each of the counties or districts would be entitled. This election was to be held no less than sixty days after the issuance of a proclamation informing the counties as to their apportionment in the first legislature.

Accordingly, on November 28, 1853, Governor Stevens proclaimed Monday, January 30, 1854, as election day. The polls were directed to be open from 9 a.m. to 6 p.m. Thurston County polling sites included Olympia, Chamber's Prairie and Ford's (Prairie). The Council or Senate comprised fourteen members, including two from Thurston County.

Olympia Whig Convention

In a January 10, 1854, letter Allen described Olympia's reaction to the Governor's proclamation and how Allen himself came to "dabble" in politics.

The arrival of the Governor—the approaching of the time, in consequence, of an organization of their government—the election of Delegate to Congress, &c.—threw the hitherto calm and quiescent people into a high state of excitement. Party lines began to be strictly drawn—caucuses and calls for conventions seemed to be the order of the day.

There was a good deal of speculation as to the relative strength of the respective parties, when arrayed against each other, and a prevailing desire to ascertain each man's views. No one, however humble his estate, but was called upon for an expression of his political views and opinions. Chancing to attend a sitting of the Whig Convention, I was called on to define *my* position, the doing of which would involve the achievement also of a speech.[1]

Allen Gives a Speech

Now, I am no orator. Nature, in furnishing me with whatever gifts I may possess, to enable me to act a man's part in the great drama of Life, denied me the faculty of expressing myself, fluently, by "word of mouth." In vain have I endeavored, at times, to follow Hamlet's advice to the players, to allow my words and sentences to go "trippingly on the tongue." Conscious of this disadvantage, on rising to comply with the request of the convention, I felt not a little embarrassed. It was a convention composed of rather more than the ordinary mental caliber. There were individuals seated there who had borne an honorable part in the forming of state and territorial constitutions, ex-honorable M.C.'s [Members of Congress] &c., and it seemed as if all at once, I was seized with a sudden sensation of choking in my throat, and a thirst that nothing could allay. But there was no way but to burst in—the difficulty being to make a commencement.

Recalling that other epoch in my life, when a member of the convention assembled to memorialize

Congress to give us a separate organization from Oregon,[2] and when, awed by the solemn responsibility resting on me to demonstrate to that convention the momentous interests entrusted to their deliberations, and the dangers to avoid—entering into a retrospective view of all the nations and empires that had ever flourished—I was so summarily "choked off," for taking a too comprehensive scope for their limited time, it seemed prudent on this occasion to commence with a *brief* preface.

The old prelude of school-boy days came to my mind—"You'd scarcely expect one of my age," &c.—which, however, would have been simply absurd, as two years (equal in experience to ten spent elsewhere,) had been added to my age since that memorable epoch in my life, and that they *did* expect something from "one of my age" was evident from their confounded look of expectancy, as though nothing but a dash of Demosthenes would satisfy them.

I recollected that, of all the distinguished orators, those had impressed me most who had evinced earnestness and a sincere belief in the sentiments and opinions avowed, and it seemed to me that, by a kind of mental and subtle instinct, I could detect whether the speaker was sincere, or "acting a part." I *know* I was sincere in my convictions, and, once started, could be in earnest. Without desiring to seem to wish to attach too much importance to what may seem to others a trivial occurrence, (supposing me to possess any political aspirations,) I must say it was to me an epoch, a crisis, in my life.

It was known among my neighbors, associates and friends that I was a "Free-Soiler,"[3] and cherished certain other radical and ultra opinions, not palatable to "fogies" in general, or timorous conservatives. But it was not generally known whether these opinions were mere abstract speculations or honest convictions forming a rule of life—something to shape my actions by—cherished and voted for, at all hazards and sacrifices, (if any there were.) In short, they were, so far as political opinions were concerned, in ignorance as to what manner of man I was—whether a "reed shaken by the wind," or erect and firm as one of our own "towering pines."

Mentally recalling the advice given to another unfortunate stutterer, under similar circumstances

"Orator: amid the crowd
Moved like waves at thy behest
Hearest thou that which, shouted loud,
Were a terror to thy breast—
Be firm! be true!
Then, fall what may upon thine ear,
Thy heart shall know no onward fear."[4]

I commenced, and at once avowed myself "A Whig, relatively to our pseudo Democracy"—which declaration may possibly surprise you, as, discoursing of politics in my past letters, I may have expressed my belief that from that party would proceed the measures of reform and progress Free Soilers demand from all who claim their influence and votes—which opinion at that time I firmly believed, thinking there was a charm in the same—an aroma, a latent power and savor in their ranks and past history, which would, after a time, place them in the van of the great "American party" already forming and destined to overshadow all others. Recent events, however, have bitterly disabused me of this belief, and, putting but little reliance in the *professions* of any party, I believe firmly now that it will be the Whigs who will save the country from the Slavery propagandists."[5] Such were my reasons for avowing myself a Whig, *but only subordinately*. I should vote for a Democratic Delegate to Congress as no principle is thereby compromised, and the very fact of his *being* a Democrat would give him greater influence with a Democratic administration to procure grants and appropriations, for our public works and roads. In all matter of public interest I should, in voting for candidates, vote for such men, by whatever party nominated, who would be of most service to the Territory.

Furthermore, (and *here* I became *terribly in earnest*, and was undoubtedly creating a sensation *now*—not so much from the eloquence exhibited, but from the boldness and utter "abandon" of the sentiments avowed,) I was primarily above *all* parties, cliques, caucuses—subject to and to be controlled by none. I knew not nor cared if there were any others there present in that assemblage of my views. I avowed myself indifferent to all office preferences or the dicta of majorities, knowing and feeling myself *to be right!* A Free-Soiler, "body and breeches," heart, soul and brain, my opinions constituted the engine and motives of my actions, hereditary, instinctive, and annealed by conviction—"bone of my bone and flesh of my flesh," and therefore, should the nominees of either Whigs or Democrats adopt this platform, and I be morally certain they would carry them out, I should vote for such—nay, more: If neither of the two parties should nominate candidates pledged to these views, and there should be enough of Free-Soilers in the territory to nominate from their own ranks suitable candidates, I would vote for them, though I voted "solitary and alone," and at the risk of having to hear the old taunt of "throwing a vote away"—as though a vote given for a principle, and consecrated to Freedom, *could* be ever lost!

With a few, astounding and never-before heard of tropes and figures of speech, uttered in my most boisterous tones, (considerable power having been added to my lung arrangements since I left home,) the most affecting of which was the representation of the "Eagle of Freedom swooping over that land

'Where Slavery's dark pinions gloom the hours,
And, unavenged, stalks forth dark Crime,'[6]

—seeking in vain for a spot consecrated to Freedom—and at last, soaring northward, finds Mt. Rainer, descends upon its heaven-piercing summit, and complacently looks down on Puget Sound and the good people who "thereabout do dwell." I concluded my tremendous effort with a few parting words to the astonished audience—that, if anything could be made of such an untractable and angular individual as he who had just addressed them, they were welcome to put him to use, and sat down with the comfortable and cheerful conviction that I had "done it," and was politically dead, defunct.

I was returning with this pleasant conceit in my mind, when I was arrested by the loud voice of one of the members in attendance, (a herdsman or shepherd, formerly in the employ of the Hudson Bay Company, and now an extensive cattle-raiser,) of an enthusiastic temperament, and, when excited, slightly profane, who frankly declared "he didn't know what it was all about, but he'd be d—d if he didn't like them sentiments!"

When the Convention adjourned, I was astonished to learn that there were Free Soilers sufficient to make their influence known and felt, and that they had coalesced with the Whigs and nominated an equal ticket, on which, to represent the Free-Soil sentiment, for the Council or Upper House, I was placed, with one of the oldest settlers, the most wealthy and influential in the territory, and my friend, Judge Yantis. Rayther flattering wan't it?—one of the few instances where Virtue is not merely "its *own* reward!" There's egotism for you![7]

The *Washington Pioneer* on December 31, 1853 reported in fewer words, "The Whig party of Thurston county held a convention at Olympia, on Saturday last and nominated the following ticket for members of the legislative assembly: For the Council—B. F. Yantis and E.J. Allen."[8]

Allen's January 10, 1854 letter continues:

I feel perfectly indifferent as to the result to speak candidly. I am hardly patriotic enough to care about serving my country four months for $3.50 per day—mere laborer's wages here—when there is no great principle at stake. There will be enough "good men and true" elected to see that our constitution will be preserved from all savor of slavery, and as for voting to test the respective strength of the parties, our public interests are common to all, and local questions will concern only the site of the future seat of government, opposing interests of diverse routes for new roads, &c. The Delegate to Congress will doubtless be a Democrat, voted for alike by both parties for reasons before stated.

I shall not interest myself personally for my election. I would rather be untramelled and not forego the probability of being appointed as clerk of code commissioners, or some other situation that pays twice as well as legislating.

McClellan Endorses Allen

In the 1913 edition of *The Canoe and The Saddle*, Allen wrote about Captain McClellan's view of the upcoming election: "In the fall of '53, the district was about to poll its first vote. It was overwhelmingly Democratic. Some enthusiastic friends set up a Whig ticket on which I was one of the two candidates for the territorial Senate. Capt. McClellan told me that he had never voted. If he was anything, he said he was a Democrat, but he was going to vote for me."[9]

Governor Stevens announced the final results.

PROCLAMATION
In accordance with the act of Congress approved March 2d, 1853, entitled "an act to establish the Territorial Government of Washington," I, Isaac I. Stevens, Governor of the Territory of Washington, do declare that according to the returns of the election held in the Territory of Washington, on the 30th day of January, 1854, in conformity with the Proclamation bearing date November 28th, 1853, for the election of delegate to Congress and for members of the Council and House of Representatives of the First legislative Assembly declare…In the election of two members of the Council from the county of Thurston, B. F. Yantis received 214 votes, D. R. Bigelow 198 votes, S. D. Ruddell 163 votes, E. J. Allen 148 votes, Samuel James 1 vote and O. Cushman 1 vote. B. F. Yantis and D. R. Bigelow are therefore duly elected.[10]

Edward Jay Allen's political career was over.

Back to the Cascades

A Pause in the Letters

The Yale scrapbook contains only one letter for the year 1854. Rebecca Allen's correspondence suggested that at this point Edward stopped the publication of his letters. She hinted that he found some fault with them, but never fully explained what that fault was. Clearly the decision disappointed many people. In a January 31, 1854, letter Rebecca vented her frustration over Edward's action to her brother William:

Several persons have asked if we have heard from Eddie lately, when we say yes, they ask if we are going to have them published, when we tell them what Eddie says, they seem surprised and say that Eddie is mistaken. The clerk at Jones and Quigg's asked Father if we were not going to publish, says he sent all his papers to New York, where Edward Allen's name was as familiar as household words. Jimmy McDonald, whom you know is not very enthusiastic on such matters, regrets it much, says he will write to Eddie about it, told us a man in Ohio who took the *Dispatch*, had written to him to know if the Oregon Trail was going to be continued; if not he would give up the paper, as he only took it for the sake of Eddie's letters.

This man is a son of Mr. Kennedy. Mr. Morrow came in to settle the tax a few days since, asked if we had heard from Eddie, and when were we going to have it published, regretted much to hear Eddie's regards on this subject, said the public were much interested in them, and were on the look out for more. One man came on business, and I never heard anyone speak as he did. After going on at great length, saying he [Eddie] would be governor, a member of Congress, President of the United States, finally said with such a young man as that, the government would not know what to do with him. After he was gone Father dryly remarked the government would be greatly put to it, the constitution would have to be altered! Mr. Forsyth of Enon valley spent one night with us a week ago or so. Says Eddie is going to be governor of Washington. Says there used to always be a dispute when the *Dispatch* would come in the morning, as to who should read it first.[1]

On February 6, 1854 Rebecca wrote to Edward:

And let me add, dear Eddie, you underrate those letters. If you could only hear the spontaneous compliments passed upon them you would perhaps estimate them more highly. I have in former letters told you much on this subject. Many persons have enquired whether you are going to write any more. Some are going to give up the paper if the Oregon correspondence is suspended. What hath been said and by whom I will leave Amelia to tell. It will be a delightful task for her to account the praises of our brother.[2]

Unfortunately Edward stuck to his decision and did not resume writing for the *Dispatch* until 1855 when he recounted the story of his return to Pittsburgh. The action to cease publication was also a disappointment to historians, for 1854 was a very busy year in the life of Edward Jay Allen.

Contract Awarded

Allen's January 10, 1854 letter struck a hopeful note regarding the military road project and continued with praise for the new governor.

Both the Governor and Capt. M'C. have spoken to me about taking the contract for converting our emigrant road into a military road. The Governor, in his social intercourse with his fellow citizens, is remarkably courteous and genial. He has brought with him the finest library in the territory, the use of which he has kindly tendered me. The people of Washington

Territory have reason to be proud of their Governor. He seemed, from the very moment he entered the territory to identify himself with her interests.[3]

The letter of January 10 is the only 1854 letter in Yale University's Allen scrapbook. For the remainder of this year news of Allen's activities came from other sources. Allen himself later recounted that he served as Captain McClellan's secretary during January, February, and early March of 1854.[4] His duties most likely included helping McClellan write reports to the War Department about his exploration of the Cascades. George Gibbs, sharing the cabin with Allen and McClellan that winter, was also writing formal reports of his exploration of the overland route to Shoalwater Bay (today's Willapa Bay) for both the governor and the War Department, as well as compiling his Chinook jargon dictionary. Perhaps Allen also helped Gibbs with these tasks. Allen's obligations to McClellan would have ended March 6 when Governor Stevens relieved the captain of his duties in Washington Territory and ordered him to report to the other Washington for further assignment.

Stories in the *Pioneer and Democrat* and other papers revealed that Allen was indeed awarded the contract to upgrade the People's Road. The *Pittsburg Daily Dispatch* reported:

FROM WASHINGTON TERRITORY—We have just received a copy of the Olympia *Pioneer and Democrat*, of March 11th. In the proceedings of the House of Representatives of Washington Territory we notice the introduction of a bill to authorize Edward J. Allen, (our well-known correspondent,) "to establish and keep a ferry across the Columbia river near Fort Walla Walla, at the point where the emigrant road from the States to Washington Territory crosses said river." The bill makes it Allen's duty to have a suitable ferry boat and allows a charge of $6 per wagon, and fifty cents per head of cattle for crossing. An amendment was carried to add a charge of fifty cents for every passenger not accompanied by a wagon. The bill then passed the House. In the Senate, however, we are sorry to see the bill indefinitely postponed. Sorry to see this, Ed. "Better luck next time." In the same paper we find a message from Gov. Stevens which states that: In answer to the resolution of the Council asking for information in relation to the military road from Walla-walla to Steilacoom, I have to state that fifteen thousand dollars remain unexpended—that a contract, subject to the approval the Secretary of War, has been made with E. J. Allen, Esq., for continuing the work on the road, and that the Secretary of War has been urged to ask Congress to appropriate twenty-five thousand dollars to complete the road, including the payment of the work done by the citizens of the Territory for the past year.[5]

Shirley Ensign, whom Allen sent to the Columbia River in 1853 to build a ferry, did not complete the work in time for the Longmire wagon train to benefit. The legislature eventually granted a ferry charter to Allen and Ensign in partnership,[6] and in September 1854 a ferry was indeed in business. Ezra Meeker used it as he traveled east to Oregon to guide his family's wagons over the Naches Pass road. Meeker stated that Shirley Ensign operated the ferry.[7] Winfield Scott Ebey also used the ferry that September and noted it in his diary.

Lt. Arnold Recommends Naches Pass for Military Road

First Lieutenant Richard Arnold, who had accompanied Governor Stevens west as a member of his railroad survey team, was officially tasked with completing the military road.[8] Allen, a civilian, was awarded the contract but the military, in the person of Lieutenant Arnold, controlled the distribution of funds. On May 18 Arnold, Allen, and company left Puget Sound on a six-week exploring trip "to survey the best military route between the Yakima valley—the eastern slope, to the west," assessing the snow conditions and the suitability of Naches Pass as a railroad route—a project that Captain McClellan failed to do.[9] Arnold reported four feet of snow on the ground for five miles, and in many places from six to ten feet, but recommended that this route be officially (finally) designated as the military road. Arnold held off on the railroad recommendation.

Gibbs and Arnold Meet Allen

On July 2, 1854, Lieutenant Arnold made a second exploratory trip to Naches Pass, this time accompanied by George Gibbs and a supply train headed by an unnamed white man and two Indians.

They closely examined the 1853 road and recommended improvements that were sent back to Allen's work crew by courier. Gibbs kept an unpublished journal during that trip, which opens a window into the activities of Allen and others traveling over Naches Pass that summer.[10] Gibbs stated that Allen's route down to the White River crossing was too much to the left and therefore longer than it needed to be.[11] His July 5 entry said, "Mr. Moore went back with directions to Mr. Allen for the guidance of his working party—further instructions to be sent back to Porters[12] on the next opportunity." Gibbs commented on Porter in the July 4 journal entry: "Porter has a considerable number of cattle which wintered in the Yakima country & were brought over this year." This suggests that Porter used the Naches Pass wagon road constructed in 1853 to drive cattle to the Yakima country in the fall and that he was able to bring them back by July. Such an event raises interesting questions, such as how many cattle Porter drove over Naches Pass, when he drove them, and the quality of the road.

On July 5 at the base of Mud Mountain, Gibbs "found here a party of miners encamped on their way across the mountain, one of the party having cut his foot with an axe." Just east of Mud Mountain Gibbs noted, "Here we met others of the same party of miners encamped." The next day (July 6), "After getting breakfast we pushed on ~~by the trail~~& during the morning met father Pandosy of the Atanam mission[13] with a large band of Yakima Indians on his way to Nisqually. Stopping to advise with him about a guide acquainted with the passes of the Cascades he recommended an Indian named tau-nou-ne-moh, residing either at Kititas or Selah."

Gibbs and Lieutenant Arnold then "[c]rossed Green River a little above its mouth and camped that night on Bare prairie the 'Little prairie' of Wilkes, which lies between it and White river. The prairie is covered with uva ursi[14] and the little grass around is eaten off by the Indian horses. We found some and camped here & others came in during the evening among them the son of Owhei, & his wife, whom we met the last summer."

The next day the two explorers climbed to the summit of LeTete[15] in an attempt to obtain views of a second pass of which they had received reports. The view was intriguing but inconclusive. On July 8 the party started up the Green River. "About half way to the foot of the first hill cross a branch coming in from the right half the size of the stream. Apparently Mr. Allen suppose this to be a ~~bran~~ part of the main river forming an island but it is not. It is probably formed by the union of the streams from the ridge we afterwards ascended."[16] The next

Meadow near the summit of Naches Pass. Photograph by Dennis Larsen.

two days (July 9-10) were spent in search of the suspected pass. Arnold and Gibbs ascended the Greenwater River, climbed the ridge leading to today's Noble Knob and concluded that there was no such pass.

Accordingly on July 10 they dropped down to the emigrant road and "ascended the mountain to examine the approaches to the well known hill, the last descent of the Emigrants from the summit." Of Government Meadow Gibbs wrote, "One of the emigrants of last year informed us that on the largest prairie was a spring hole some yards across, into which they had stuck a pole 50 or 60 feet long, without finding bottom."[17] Gibbs described the summit prairies as very marshy and suggested draining them with ditches as part of the road building effort.

The explorers went just past Edgar Rock on the east side of the pass and upon returning on July 15 "travelled to the first prairie on the summit, the ascent of the mountain taking two hours. In the afternoon it commenced raining & some snow fell with it. We here met Mr. Hurd who was going over to meet the emigrants at Fort Boise & who brought letters for Lt. Arnold & news that a great excitement prevailed in the settlements about alleged gold discoveries in the Yakima river. As we had met in going over a Mr. French, one of two had been prospecting there this summer & who was packing back his tools, we were not a little surprised."[18]

On July 17 they descended Mud Mountain and "[t]ravelled to first prairie beyond these mountains & camped. We met Mr. Allen's working party a little before we reached it, in consequence of which Lt. Arnold concluded to remain over." The next day "[r]emained at this camp. Lt. A. being busy in directing the work." On July 19 the explorers left Allen and his work crew and returned to Fort Steilacoom "by a forced march."

The *Pioneer and Democrat* reported on July 8 that Allen had just visited Olympia and was on his way to the mountains again.

> The road across the Cascade mountains, for the construction of which our citizens subscribed so liberally last year, and, which was opened sufficiently to admit the ingress of a portion of the immigration of last year, will soon be rendered eminently practicable as a good wagon road.
>
> The (Nah Chess,) pass through which the road has been constructed, has been decided upon by the Secretary of War as the one for the construction of the Military road leading from Fort Walla Walla to Fort Steilacoom, for which an appropriation was granted by Congress of $20,000—$15,000 of which will be expended on it this year.
>
> Mr. E. J. Allen of this place, to whose energy and the faithful discharge of duty of the men under him, the people of the territory are indebted for the prosecution of labor on the road thus far, started from Steilacoom yesterday with a party of thirty men, to complete, as far as possible, the work commenced, in time better to enable immigrants to cross the mountains with less difficulty, and more ease and comfort than has heretofore been enjoyed for admission into either this or Oregon Territory west of the mountains. Instructions from Washington with regard to the expenditure of the appropriation of Congress, have been so long delayed, and the season is so far advanced, that it will require all the energy of Mr. Allen and faithful labor of his party to be of material service for the benefit of the present immigration; but judging from the past we have reason to expect much as the result of the labor about to be undertaken.
>
> It is not at all improbable that Nah Chess pass will be found to be a better grade than any other pass of the Cascades leading into this territory; and the better to decide the matter, Lieut. ARNOLD, attached to GOV. STEVENS' expedition is at present engaged with a small party in making a thorough examination of the pass to determine its practicability for the construction of the Pacific railroad through it, whenever that great national work is undertaken. At all events a good wagon road can be made in that direction, and we expect it will be done this season, as far as $15,000 can make it.
>
> In this connection we would refer to the fact that WHITFIELD KIRTLEY, ESQ., JAMES K. HURD, and many others of this place and vicinity, are entitled to no small praise for the part they performed in the competion [completion] of the road thus far. We understand that Mr. HURD starts for Ft. Boise on Monday [July 10],[19] to apprise the immigrants of the road being opened.[20]

Allen's activities were next mentioned in August:

> Messrs. Ensign, Kirtley and Blankenship, of this place, returned on Saturday last [August 12] from a gold prospecting tour on the eastern slope of the Cascade mountains and seem satisfied of the existence of gold in the country traversed by them, but express a doubt as to its existence in sufficient quantities as to justify extensive operations being entered into in consideration of the present prices paid for labor.
>
> They report favorably of progress on the immigrant road, upon which Mr. E. J. Allen and a party of men are engaged in its more thorough construction, and express the belief that the expectations of the immigrant will not be disappointed in a transit across the mountains. The Indians reported the arrival of three wagons at Fort Walla Walla.[21]

Climbing Mt. Adams

That summer Allen added another page to his fast-growing resumé: he led the first known party to climb Mt. Adams. George Himes, Naches Pass pioneer of 1853, and who by the turn of the century had become Secretary of the Oregon Historical Society, wrote an article for *Steel Points*, the journal of the Oregon Mazamas climbing club, in April 1907 telling about early climbs of Mount Adams. In the July 1907 issue he added:

As one of the results of the thirty-fifth annual meeting of the Oregon Pioneer Association, I secured a little data respecting the climbing of Mt. Adams, in which you will doubtless be interested, and which has never before been made a matter of record. In conversation with Mr. A. G. Aiken, of Marshfield, Coos County, Oregon (who, by the way, was in the same train to which I belonged while crossing the plains in 1853), I learned that Edward J. Allen, Andrew J. Burge, and himself made the ascent late in August or early September, 1854. These persons belonged to a party of men who left Steilacoom a few weeks before the ascent was made to work on a military road that was then being constructed by government authority from the Columbia River through the Naches Pass to Puget Sound, following generally the trail partially made by the immigrant party of 1853. It was while this company was camped a few miles northeast of the base of Mount Adams that the three above-mentioned persons decided to make the climb. As I have had a personal acquaintance with all three men, I have no doubt as to the fact of their making a successful ascent. At last accounts Mr. Burge and Mr. Allen were living, and as it is a matter of general interest, to all Mazamas at least, I hope to secure full details of this, probably the first known ascent of Mt. Adams, in a few weeks.[22]

Aiken himself gave a few more details in a 1909 interview:

A. G. Aiken, for many years a resident of Marshfield, and one of the pioneers of Coos County, tells of making an ascent of Mount Adams in 1854, which he believes was the first ascent of that mountain ever made; at least there were no traces of anybody having climbed it before that time. Since others claim to have made the first ascent in 1866, Mr. Aiken recalls the incidents of the expedition:

I was working on the military road being built to Fort Walla Walla. E. J. Allen was superintendent of the work, and it was he who made up the party to explore the mountain. Allen, Andrew Burge, myself and one other man whose name I cannot recall made the trip. Others in the road-building corps could have gone had they cared to, but as I remember, only four of us went. It took us a day to reach the foot of the mountain and we camped there for the night. The next day we reached the summit. That night we camped at a spring near the summit. On the top of the mountain we made a pile of rocks and put up an American flag, which Allen had brought with him. I wonder now that the party that went up later did not see the pile of rocks we left. The view from the summit was beautiful. We could see Mt. St. Helens, Mt. Rainier and Mt. Hood. I do not recall ever enjoying a more beautiful view than was before us at the top of that mountain. As I remember it now the climb was not a particularly difficult one. Fifty-five years has passed since then and I am not sure of the exact date, but I think it was in the latter part of September or possibly the first part of October, 1854.

Mr. Aiken came to the Pacific Coast from Pittsburgh in 1853, and shortly after he climbed Mt. Adams he came to Coos County, and has resided in this locality ever since.[23]

Fred Becky, the preeminent authority on the climbing history of the Cascades, agreed with Himes and credited Allen, Burge, and Aiken with the first ascent of Mt. Adams via the north ridge and suggested that Colonel B. F. Shaw was probably the fourth person mentioned by Aiken.[24]

An Impossible Feat

When Allen wrote for the 1913 reprint of *The Canoe and the Saddle*, he left a somewhat puzzling account of his summer's work.

I am compelled to say I do not think McClellan's survey of the Cascades for a railway route was very thorough. When after one summer of volunteer work by the citizens endeavoring to make a passable road through the Nahchess Pass, Capt. McClellan gave me the contract to expend what remained of the $20,000 appropriated by Congress for that purpose, he suggested that he could make an engineers examination of the Pass. I replied that such an examination would exhaust the whole amount, and then would only demonstrate that it would require at least $500,000 to construct what would be but a faint approach to a "Military Road." Hence that idea was abandoned.

I found that $5,000 of that amount had been expended in his general examination of the Cascade Range, leaving but $15,000.

Lieut. Arnold, of what was then called the Dragoons, was detailed to go over the route with me. I think the amount of $15,000 and a passable road for about 135 miles did not seem to him to have any close connection with each other. I really forget whether McClellan was of the party, but think not at that time. He, however, came later when we had reached seeming 'impasse, where the open, if rough, Pass ended, as all Cascade passes do, in an abrupt mountain closing up the gap. He said I had done well with what I had expended, but of course I could do nothing to overcome this obstacle.[25] To which, in the heat of youth and with some idea of what would be deemed possible by an emigrant that would seem besotted ignorance to an engineer, I replied: "I will make up that almost perpendicular 1,200 feet not only a road that an immigrant can get down, but one that six yoke of cattle can haul 1,000 pounds up." To which he gave a kindly but incredulous shrug of the shoulders. My difficulty, of course, was not an engineering one, but simply a matter of finances. I had but a few thousand dollars left. When he came back later, at my request, we had constructed a road up which I hauled, with four oxen 1,500 lbs. It was buttressed up an average of fifteen feet, and in some places forty feet, with the huge trees that covered the mountainside and was stayed down the mountain, from tree to tree, with thousands of braces. It was impossible, but we were ignorant, and not fully conscious of this impossibility, so we did it." McClellan stood on the highest point of the buttresses and said, "Young man, do you know what you have done here? Under the conditions, Napoleon's passage of the Simplon was an engineering feat no greater than this."

I wonder how much of that road exists to-day. It did not seem to me very extraordinary then, but I was only twenty-two. I am now in my eighty-fourth year, and have learned that youth and its inability to recognize obstacle are great factors to success. Some large measure of McClellan's opinion of our work in the Nahchess Pass went into his report. I remember looking over a map of his reconnaissance of the cascades, and noticed a camp designated 'Hellis-del-ight.' I could not recognize any Indian dialect in this nomenclature, but McClellan explained that the camp was an unusually unpleasant one, and that the name was a disguise for 'Hell's Delight.' I presume that map is on record in the War Department. It is a testimonial at least to McClellan's sense of humor.[26]

Allen's account of McClellan making his way up to see the completion of the work on the "cliff" at the end of summer cannot be correct. McClellan left the territory in the spring of 1854, spent the summer on assignment in the West Indies, and was in Washington, DC, during the fall and winter of 1854-55. There is no evidence that McClellan ever saw this part of the road. Some sixty years later Allen was possibly confusing memories of Lieutenant Arnold and Captain McClellan. If these 1854 events occurred, it may have been Lieutenant Arnold who stood at the top of the cliff when Allen made his bold prediction—one that came to fruition, for he did indeed build the road that some military officer praised.

Meeker Describes the Trail

Ezra Meeker, who made a round trip over this road in September and October 1854, noted the work that Allen's crew had done, but he failed to encounter any of the workers. Perhaps this was because he stuck primarily to the old Indian trail on his way out, which at times diverged from the wagon road, and by the time of his return trip the road workers had finished for the season. Or Meeker may have taken Mike Simmons' old 1850 road to Allen Porter's residence on the White River and missed the road crew entirely because they were working on the section of the new road which bypassed Simmons' previous effort. Meeker described Allen's road:

And then, the road. Such a road, if it could be called a road. Curiously enough, the heavier the standing timber, the easier it had been to slip through with wagons, there being but little undecayed or downed timber. In the ancient days, however, great giants had been uprooted lifting considerable earth with the upturned roots, that, as time went on and the roots decayed, formed mounds two, three, or four feet high, leaving a corresponding hollow in which one would plunge, the whole being covered by a dense short, evergreen growth, completely hiding from view the unevenness of the ground. Over these hollows the immigrants had rolled their wagon wheels, and over the large roots of the fir, often as big as one's body and nearly all of them on top of the ground…When the timber burns were encountered the situation was worse. Often the remains of timber would be piled in such confusion that sometimes wagons could pass under the logs that rested on others, then again, others were encountered half buried, while still others would rest a foot or so from the ground, these, let the reader remember were five feet or more in diameter, with trunks from two to three hundred feet in length. All sorts of devices had been resorted to in order to overcome these obstructions. In many cases, where not too large, cuts had been taken out, while in other places the large timber had been bridged up to by piling smaller logs, rotten chunks, brush, or earth, so the wheels of the wagon could be rolled up over the body of the tree. Usually three notches would be cut on top of the log, two for the wheels and one for the reach or coupling pole to pass through…Small wonder that the immigrants of the previous year should report that they had to cut their way through the timber, while the citizen road workers had reported that the road was opened, and small wonder that the prospect of the road should have as chilling effect on my mind as the chill mountain air had on my body.[27]

Meeker described emigrant ingenuity in the face of challenge:

One day we encountered, a new fallen tree, as one of the men said, a whopper, cocked up on its own upturned roots, four feet from the ground. Go around it we could not; to cut it seemed an endless task with our dulled flimsy saw. Dig down boys said the father, and in short order every available shovel was out of the wagons and into willing hands, with others standing by to take their turn. In a short time the way was open fully four feet deep, and oxen and wagons passed through under the obstruction.[28]

The *Pioneer and Democrat* reported that Mr. Silas Galliher, the advance guard of the first of the seven wagon trains that would come over the Naches Pass wagon road, had arrived in Olympia on August 29. The newspaper stated that Galliher "reports the road over the mountains as being much improved from what it was represented to be last year, and that the party under Mr. E. J. Allen are within about 18 miles distant on the western slope of the summit, proceeding as rapidly as circumstances will warrant eastward."[29] Interestingly, in that first wagon train which reached Olympia on September 14 was James Kirtley, brother of Whitfield Kirtley.[30]

A Well-Used Tent

The 1854 Ebey wagon train provided the last contemporary account of Allen's activities that summer. Winfield Scott Ebey, age twenty-two, led the wagon train and kept a diary. On September 25 he noted that they had encountered Edward J. Allen's thirty-man road construction crew just east of the summit, spent the night with them and that he had sold Allen his well-used tent. The original purchase price was three dollars. The sale price to Allen was ten dollars. Ebey chuckled over the sale. "[T]his evening disposed of our faithful old tent to Mr. Allen…we had used it pretty well up…"[31] Anna Marie Goodell who was in the wagon train also kept a diary. She stated simply, "We traveled 8 miles & camped where the men were at work on the road."[32] On September 26 the Ebey train moved on to Government Meadow. In his diary that night Winfield Ebey wrote, "Passed over a part of the New road opened by Mr. Allen It is well cut out but the ground is rough & in many places Swampy."[33] The next day as the wagon train descended the mountains Ebey noted, "[T]o our right was 'Mount Ike' a tall bare rocky Peak[.][34] On its Summit we could just See a Flag waving in the morning breeze[.] It was put there a few days since by Mr Allen."[35] He also described Allen's work further down the trail. "Last year [1853] the emigrants were compelled to let down their wagons down by ropes or with large trees tied to them. Here we found that Mr. Allen had made a fine grade. So that we descended quite easily. For about a half a mile the road is cut into the side of the mountain. On the lower side is a hand railing for the entire distance."[36] On September 30 they broke out of the forest and camped in a grassy valley just northeast of the present-day city of Buckley.[37]

Mystery Voyage on the *Major Tompkins*

The last news of Allen in 1854 came on December 23 when the *Pioneer and Democrat* reported the arrival of the steamer *Major Tompkins* from Victoria and waypoints with a large number of passengers, among whom was listed E. J. Allen.[38]

Allen's activities in the fall of 1854 are a blank. Blanche Billings Mahlberg wrote in her 1953 *Pacific Northwest Quarterly* article, "He went to San Francisco and contracted to supply timber for construction in that fast growing city."[39] Allen's 1853 letters mentioned making such a contract which might explain why he was on the steamer *Major Tompkins* that December. When writing her article, Billings had access to documents the authors have been unable to locate.[40] Thus a late fall trip to San Francisco cannot be ruled out.

However, Allen might have simply gone to Steilacoom for a final visit with friends and acquaintances in that city before returning to his cabin in Olympia from which he would soon depart for Pittsburgh and home.

San Francisco, view looking toward the bay, circa 1850, by Frank Marryat. Library of Congress Prints and Photographs Division, Item #92522389.

Homeward Bound

It was time for Allen to go home. Before leaving Pittsburgh in 1852, family letters had referred to a period of two years allotted for his adventure. That time frame had already expired, he had proved up on his donation land claim, and he had clearly regained his health, his primary reason for coming west in the first place. After a stay filled with adventure and accomplishments, he returned to the East Coast via the ocean route, rather than overland. In the last of his letters to be published in the *Pittsburg Dispatch*, Allen told of his return voyages.

Farewell to Olympia

AT SEA—HOMEWARD BOUND—*Steamer Uncle Sam*, March 1854[1]—*Messrs. Editors:*—On my way homeward, I write you my last letter dated from the Pacific—the anticipations of home and home friends not entirely banishing regret at leaving the little cabin by the seashore where for many months time has dealt so gently with me that its ebbing was almost unnoted. On the 11th of January[2] I sat for the last time upon the little stoop of the cabin, looking out over the calm waters of the bay, and remembering me of the evening nearly two years agone when I first floated over in my little canoe and took possession of the claim which has ever since been home to me—a home so dear that I reproached myself as I almost imagined that the road to my old home led not into light but into shadow; wondering what the fates had yet in store for me, whether my lines would ever tend hitherward again, or whether I was to sink again into the current of city life, floating as a unit on its broad tide, and remembering as in a dream the escapade of three years which formed the one episode of a life time. Those who had been with me when I first sat under the shelter of my little cabin had wandered on diverging paths; one homeward to Ohio, another to the mines of Australia, a third to the Southern Vallies of Oregon, a fourth about seeking a home upon the lands of the Nez-Perce. I bidding farewell to the spot from whence all our pathways radiated, never again in this life to converge together. Long that quiet association had endeared to me all things animate and inanimate—the very sound of the waters upon the beach, the rustling of the cedars, the whistling of the eagle, which long toleration had so emboldened that his keen eye no longer kept note of my movements, as he sunned himself upon the overhanging cedar. My noisy friends the crows had for some reason unknown deserted me, but in their stead were crested jay birds, who disputed the scraps which I threw out, with a bright eyed pine squirrel, who had grown so bold that he not unfrequently came quite into the house. The little woodmice had since the departure of my feline friend grown "unco" familiar, but as they *would* "prospect" the water bucket, whence I would extricate them in the morning all swollen and lifeless, I was obliged by cunningly devised snares to rid myself of them—and the place thereof "knew them no longer."

The little space about my cabin had altered somewhat. The bright sun streamed in where aforetime the heavy cedars had cast shadows upon the ground—but out upon the broad bay all was unchanged, and the mellow golden light rested upon Rainer as gloriously as when upon the "evening of the first day we leaned upon our axes and drank in the "outlook" before us.

Just below me on the bay a settler[3] had taken up his claim some year ago—a singularly unenergetic fellow—who had married upon the plains and settled upon the land seemingly with no other thought than that of raising potatoes and digging clams enough for the coming meal.

—I strolled down the beach to bid them good by—found the man not at home—and stopped for a few moments to present some trifling little presents to his wife and pay her for some clearing her husband had done for me.—Talking about Olympia I asked her if she ever went to church there. She said no. She used to go to church very regularly, but Mr. Buntrager did not believe in religion, and she didn't go anymore, for she "felt as though she didn't want to go to heaven unless Mr. Buntrager he could go too."

The involuntary smile that flitted over my face died away in a vague feeling of respect for the man who could inspire a woman with such a fervent love, and as I recalled the homely features of Buntrager they grew radiant with the halo of love that was cast around them, and noble in its inspiration. The very air seemed brighter, and as I walked slowly up the beach, the waters seemed to ripple with an added melody. I murmured to myself a half-forgotten quotation. In the deep stillness of the night Claarchen prayed for Egmont, and because prayers would not save him, she died with him.[4] Half an hour afterwards I was afloat in my little canoe paddling slowly up to Olympia. A harassing time settling up my business—a good bye to warm friends—and I was embarked upon the Halcyon,[5] streaming down the Sound, outward bound.

A Rough Voyage

Three days and we were out of the straits of San Juan, gazing at the pillar of De Fuca—marking the twelve mile entrance of the Straits—a fair breeze and speeding merrily onward. So far so good—but the gale increased, and the beauty and the grandeur of the scene died away in a "sense of gone-ness at the pit of the stomach" as I leaned over the rail. All night I kept it up, with an industry worthy of better results, and morning found me in an imbecile and limp state, still going through the motions, and vainly trying to convince myself that I had "come out jolly under the circumstances" and had abated not a jot of spirit.

Sea still raging—Captain suggested eating a little breakfast—the bare thought of which sent me out over the rail shoeless and coatless, in an ineffectual endeavor to get rid of something which the "vomito" would persist in thinking was still in me. The hirsute mate suggested drinking salt water. I drink a glass out of a bucket which he dipped up for me; but he laughs scornfully and says I must drink a gallon. I dipt up frantically and tost it down, glass after glass, until, "blu-ught," up comes everything, and I collapse. No better. The man at the wheel suggests a piece of fat pork tied to a string, swallowed, pulled up, and the operation repeated until I grow better. At which suggestion I go off in a perfect torrent of retchings, subside a little, and grope my way into the cabin, barely dodge a plate of scouse[6] which the steward shies at me as the vessel lurches, but am somewhat consoled at seeing a passenger who never gets sea sick, emptying the coffee out of his shoes.

Think it will never do "to give it up so," and go up on deck to "exercise," vessel pitches so that I cannot keep my feet—old sailor plays off some little joke having reference to "sea legs." Try again, in common with passenger who "never gets sea sick," to "throw up," but fail, having no "capital to work upon;" turn round in time to catch steward with his tongue in his cheek and winking at a sailor; try to smile feebly and look as if he was laughing at the other man, but from want of "vim" it is rather a weak attempt. Head feels as though it had a tight band around it, and "keeps getting no better very fast." Don't eat a bite of anything for nearly four days, when I get so I can "peck" a little, then get worse again—and so relapse and recover a little for the whole run of twenty-five days, and at last enter the Golden Gate and commemorate the occasion by depositing my dinner within its boundary. Wind lubs—heavy sea rises, and just at dark we are sweeping in toward a point of rock where the surf beats with a noise like thunder. Captain gets excited. Man at the wheel tells me that there have been several vessels wrecked there, one a fine clipper ship, within a few months; drift down to within 300 yards and drop the anchor; drag the anchor and keep drifting in; begin to calculate chances and find them pretty slim—when suddenly a wave washes over the whole vessel and the next follows suit, and we find the anchor has reached holding ground.

Don't feel under the circumstances very much inclined to sleep, and sit up all night; just before break of day turn out and help heave the anchor; breeze has sprung up, and we glide slowly into the beautiful harbor of San Francisco. Here we find anchored a perfect fleet of vessels—among others an English man-of-war. As we pass in, the steamer "Golden Age"[7] steams out; so by two hours delay I lose a week's time in getting home. Pass innumerable old hulks anchored out—make fast to the dock, and exultant find ourselves ashore. Find to my great surprise that I am very weak, and possess a certain quaver of the leg which is not suggestive of much endurance; manage to walk however some five or six miles; eat lunch; walk again, and come into my hotel at night, perfectly broken down, but feeling brave in anticipation of the morning.

A Night on the Town and Heading Home

Next evening, I went to the Metropolitan Theatre,[8] to see the opera of L'Elisir d'Amore,[9] with Madame Bishop[10] as prima donna. Tolerably fine theatre, with but few occupants—Bochsa[11] as leader of the orchestra; Madame Bishop singing well, but I am rather impressed with the idea that the opera contains more noise than melody. . . .[12] As it appears to be the general impression that we have not got the worth of our money, we stamp away, and directly the curtain shakes, and forth comes the "Herr" and Anna Bishop—bob gravely, once, twice, thrice, and exit. As this is all, we crowd out, feeling somewhat bewildered and rather deaf; so much so that we find it necessary to get a small boy who accosts us in the vestibule to repeat his remark, and finding it relates to an investment in "popped corn," respectfully but firmly decline. A square or two in the cool night air makes us all right again, and hearing a plaintive melody stealing up the stairway of a basement, over which is pendant "Lager Bier," we go down, and hear four Germans, (one of whom is Simmons, the leading violinist of San Francisco,) play and sing together; and ensconced behind a huge mug of lager bier, flanked by a "cruller" and a segment of sausage, we sit and listen to the Marseilles, Austrian national melody, anon changing to "home sweet home," and such sad, sweet airs that we are alike forgetful of the opera and our sausage, and lean upon the table with the tears starting in our eyes, and dream such visions "of youth and home," as memory has not blessed us with this many a day. From there, a ramble through the Arcade, El Dorado, and other gambling saloons; the mechanism is very simple, consisting simply of one respectable looking old man to drawl out—"games made, no more," flirt over three cards; after which, another respectable old gentleman, armed with a long rake, hauls in the loose change which any unfortunate bystander may have placed upon the baize. San Francisco is a remarkably hospitable place. Altho a stranger, I was eleven times in the kindest manner invited by young ladies to stop in and rest as I passed by, and tho impressed with their courtesy, I was constrained to refuse, on the plea of the impropriety of making an evening call on so brief an acquaintance. Seated at the window of my little room, looking out over a portion of the harbor, I read the few remaining chapters of Wearyfoot Common,[13] with its two clearly and beautifully drawn characters; and closing it, with my mind filled with its beautiful images, fell quietly asleep, and slumbered dreamlessly in my chair till morning—I feel within me reasons for being brief. After nearly swamping my little funds with Page, Bacon & Co.,[14] and running divers other little risks, a week passed away, and I embarked on the Uncle Sam[15] homeward, via Nicaragua[16] and New Orleans; whence I now write you. Will mail this via New York. Be at San Juan tomorrow. Boat pitching violently; write this in stateroom, on washstand. You can see by the writing how the boat pitches; feel qualmish; close hurriedly. Five minutes from now my head will be over the bulwarks, larboard side; places pretty well taken now. Blu-ught!
EDWARD J. ALLEN.[17]

An unattributed article in the scrapbook quoted a letter from Allen and fellow passengers:

We take great pleasure in laying before our readers the following complimentary Card to our Friend, Capt. Churchill, of the steamship Prometheus,[18] and the officers under his command.

ON BOARD STEAMSHIP PROMETHEUS,
New Orleans, March 17, 1855.
CAPT. HENRY CHURCHILL:

Dear Sir—We, the passengers on the homeward-bound passage from San Francisco, being about to separate at New Orleans, cannot leave the hospitalities of the steamer Prometheus without tendering to yourself and officers our sincere thanks for the many courtesies extended to us while on the passage—a passage so pleasant that it left us nothing to be wished for. To yourself, particularly, we would extend our warmest regards, and to those whom fortune may hereafter lead to the shores which we have left, and who wish the grand *desideratum* of gentlemanly officers and a generous table, we would most heartily recommend the steamer Prometheus.

JN. B. PRISTON,
EDWARD JAY ALLEN,
BATTES IVES,
BENJAMIN CLOSE.[19]
[And thirty others.]

The final clipping in the scrapbook is from the *Pittsburg Dispatch* editor:

Back Again.—Mr. Edward J. Allen, whose graphic letters from Oregon Territory and elsewhere, have given the readers of the *Dispatch* such satisfaction, arrived in the city last evening, per the Western cars. He looks hale and hearty, and, notwithstanding his unpleasant sensations when entering the "Golden Gate," his trip seem[s] to have agreed with him admirably.

Allen took just fifty-six days to travel from Olympia to New Orleans and probably just over two months total from Olympia to Pittsburgh—compared with the six months plus that his westward trip consumed. However, his itinerary as published in the *Dispatch* erred in a few dates. The schooner *Halcyon* left Puget Sound on January 20.[20] It arrived in San Francisco on February 16, where it met the *Golden Age* sailing out of the bay.[21] Allen's name appeared on the passenger list published in the *Daily Alta California* on February 17.[22] Allen therefore attended the opera on February 17. He then lost "a week's time" due to missing the *Golden Age*, and finally left San Francisco aboard the *Uncle Sam* on February 26.[23]

The *Uncle Sam* arrived on the Pacific coast of Nicaragua, where passengers disembarked for a trip overland to the Atlantic coast. Once there, passengers embarked on the steamship *Prometheus* for New Orleans. Upon reaching the States, Allen took an unknown ship from New Orleans to New York, where he then took the railroad ("Western cars") to Pittsburgh.

Epilogue

Allen's days as a Pacific Coast correspondent were over. From here on, his life was described primarily by other voices. The Hervey Allen Papers[1] at the University of Pittsburgh contain a sheaf of clippings relating to Allen family events from 1855 and later. These and other clippings provide details of the remainder of Allen's life.

Allen's family must have been overjoyed when he returned home from his western wanderings. Allen himself must have been ecstatic at once again living in the family home, with his beloved parents and siblings close around him, and old friends coming to call.

"The Old Playground" sheet music, published in 1855, with music by J. DeRuver and lyrics by Edward J. Allen. The dedication reads, "To William, Brother By Blood And Doubly So In Heart." Library of Congress, Washington, DC. Music for the Nation, American Sheet Music, Lyricist Allen, Edward Jay, Digital Id sm1855 581590.

In a much abbreviated form, milestones and accomplishments in Allen's later life are listed below.

1855	Allen wrote the lyrics for a sentimental song, "The Old Playground."
	Allen went into the business of selling air furnaces, ventilators, etc., with a Mr. Williams.
	Allen's sister Rebecca married Jesse Turner from Arkansas.
1856	Allen wrote the lyrics for a patriotic song, "Unfurl the Glorious Banner."
1857	Allen's beloved brother William died.
	Allen married Elizabeth Wilson Robinson.
1858–59	Allen worked as a contractor on the aqueduct in Washington City.
1859–60	Allen worked as a contractor on the construction of the Virginia Central railroad near Covington, Virginia.
1860	William Hervey Allen was born to Allen and wife Elizabeth (the couple would later have four more children).
1861	Allen, as a civilian, was detained by the Confederate Army, but was later freed through an exchange of prisoners.
	Allen re-established a friendship with George McClellan, now commander of the Army of the Potomac.
	At McClellan's request, Allen met President Lincoln.
	Allen was appointed as quartermaster for the Pennsylvania 13th Brigade.
1862	Allen recruited the 155th Regiment Pennsylvania Infantry, and was appointed Colonel in charge of the regiment.

Colonel Edward Jay Allen ca. 1862-63. From *Under the Maltese cross, Antietam to Appomattox, the loyal uprising in western Pennsylvania, 1861-1865*. Pittsburgh: The 155th Regimental Association, 1910.

1862 Allen fought at the assault on Marye's Heights during the Battle of Fredericksburg, Virginia (Allen's name is listed on a war memorial standing today near the site).

1863 Allen was honorably discharged from service due to severe rheumatism.

1866 Back home in Pittsburgh, Allen became the secretary/treasurer of the Pacific & Atlantic Telegraph Company.

1870 The U.S. census listed Allen as owning real estate worth $30,000.

About this time, Allen became the mentor of John White Alexander, who would become a well-known artist.

1885 Allen published a book, *The Walls of Sand, A True Fairy Story*.

1889 Allen was a member of the South Fork Fishing and Hunting Club, which owned a dam that burst and tragically flooded Johnstown, Pennsylvania.

Allen returned to the Northwest on a business trip.

1891 Allen returned to the Northwest again on a business trip.

1897 Allen received an honorary Master of Arts degree from the Western University of Pennsylvania.

1900 Allen self-published his volume of poetry titled *Hiou Tenas Iktah* (meaning "A Lot of Trifles" in Chinook jargon).

1914 Allen made his last will and testament.

Death of Allen

On December 26, 1915, Edward Jay Allen passed away at his home in Pittsburgh. His obituary was widely published—not just in Pittsburgh newspapers, but across the state and indeed across the country, as the New York and Olympia papers also published news of the death of an old settler (although some eastern papers mistakenly gave him credit for settling in Seattle, not Olympia).

His wife Elizabeth survived him by just five days.

Allen's obituary in Pittsburgh's *Chronicle Telegraph* noted, "Col. Allen's versatility was illustrated by his ability as a business man and his literary talent, and he not only wrote but actually lived some of the most stirring chapters in our country's history. His years were long in the land and they were years of fruitful endeavor devoted to the successful service of the nation and the community in which he lived."[2] In 1916, the Pioneers' Society of Washington memorialized Allen's time in the Northwest, and hoped that "his name will always be remembered as one of the most far-sighted and public-spirited builders of our State."[3]

In the preface to his Pittsburgh Manuscript, Allen wrote, "Time has since swept away all that was typical of this era. There can never again be another Oregon Trail stretching across the continent, nor precisely the type of men that made it memorable. These men built up the great empires of the Pacific coast. The stars they added to the blue field of our flag are enduring proof of what they did. What they experienced and endured on their way westward these letters may in some degree indicate."

Although Allen downplayed his own talents and achievements, and indeed claimed that his manuscript made "no claim to literary merit," he unquestionably made his mark on both coasts, and proved himself to be precisely the type of man who made life memorable.

Edward Jay Allen and wife Elizabeth on the beach, probably at Atlantic City; undated. Hervey Allen Collection, University of Pittsburgh.

Portrait of Edward Jay Allen by John White Alexander, Allen's protégé. The original portrait is in the Hillman Library, University of Pittsburgh. Library of Congress Digital id 3b41586u.

Notes

Introduction

1. See our companion work *Our Faces Are Westward: The 1852 Oregon Trail Journey of Edward Jay Allen* (Independence, MO: Oregon–California Trails Association, 2012). Allen's detailed account of his overland experience contains a wealth of information beyond the scope of this book, and is summarized in Chapter One.
2. The Naches Pass wagon road is currently under consideration by the U.S. Department of the Interior for inclusion in the National Historic Trails System.
3. Pittsburgh has ended with an "h" since its founding in 1758 excepting the years 1890-1911, when President Harrison established a National Board of Geographic Names that ordered, among other name changes, that all cities and towns that ended in "burgh" must henceforth drop the final "h." Pittsburgh fought this order and in 1911 the Board relented and allowed the city to revert to its original spelling. Albrecht Powell, "Pittsburgh—America's Most Misspelled City." The *Pittsburg Dispatch* chose to delete the "h."
4. Hervey Allen, Carbon Manuscript of the Oregon Trail by Col. Edward Jay Allen, University of Pittsburgh Hillman Library. [Hereafter: Pittsburgh Manuscript]
5. Edward Jay Allen to Ezra Meeker, March 25, 1908, Meeker Papers, Box 7, Folder 8C.
6. Founded in 1833, the *Knickerbocker* was a monthly literary magazine published in New York. The Allen family was acquainted with its editor, Lewis Gaylord Clark. By 1840 the magazine was considered one of the country's most influential literary publications. While primarily devoted to the fine arts, the magazine often printed news items and editorials; it was unique for its time in that it paid its contributing writers.
7. Edward Jay Allen to William Allen, October 15, 1852, Yale Scrapbook.
8. In 1855, Allen wrote the lyrics for a sentimental song titled *The Old Playground*. In 1856 his patriotic lyrics called *Unfurl the Glorious Banner* were set to music by Harry Kleber. Allen self-published a volume of poetry in 1900 titled *Hiou Tenas Iktah* (meaning *A Lot of Trifles* in Chinook jargon). He also wrote other poems and lyrics.

Chapter One: Allen's Way West

1. Edward Jay Allen to William Allen, March 27, 1853, Yale Scrapbook.
2. Based on the following two passages in Allen's correspondence: "[I]n the long years in which I have been in the warehouse I had to a great extent lost my individuality.... " Edward Jay Allen to his family, February 23, 1851, Turner Collection, Duke. In a letter written two years later, Edward talked of making good wages—"certainly more than I used to have allowed me in past times, when in the grocery connection." Edward Jay Allen to his family, February 20, 1853, Yale Scrapbook.
3. Edward Jay Allen to his family, February 23, 1851, Turner Collection, Duke.
4. Edward Jay Allen to his family, February 23, 1851, Turner Collection, Duke.
5. Amelia Allen to her family, March 27, 1851, Rosanio Collection.
6. Allen traveled west with three friends: Jacob Resser, J. William Carnahan, and William McClure.
7. Pittsburgh Manuscript: Introduction.
8. Allen and Brooks encountered each other in the fall of 1852 on the Cowlitz Trail, where Brooks informed Allen that he had been chosen as a delegate to the Monticello convention, whose purpose was to separate Washington Territory from Oregon. Allen stayed with Brooks in Olympia that winter. The story of the two men is detailed later in this work.
9. The story of the riverboat journey, and detailed account of Allen's overland adventure is chronicled in Larsen and Johnson, 29-32. The complete story of Allen's arrival and stay at Council Bluffs can be found in his Pittsburgh Manuscript, 1-6.
10. The foursome joined the Koontz family wagon train at Council Bluffs. Koontz family records indicate that there were twenty-four wagons in the train whose members were all related to each other, and that it originated from Wapello County, Iowa. Randy Brown and Reg Duffin, *Graves and Sites on the Oregon and California Trails* (Independence, MO: Oregon-California Trails Association, 1998), 89. Allen said there were thirty-two wagons in the train and that some farmers from Ohio also traveled with the Iowa contingent. Pittsburgh Manuscript: 3. The Ohio folk and Allen likely accounted for the additional eight wagons.
11. Pittsburgh Manuscript: 5.
12. Blodget's Express was founded in the spring of 1852 by George E. Blodget and R. S. Raymond of Milwaukee to deliver mail from selected sites along

the Oregon Trail such as Devil's Gate, South Pass and Goose Creek, 200 miles west of the Continental Divide. John D. Unruh, *The Plains Across* (Chicago: University of Illinois Press, 1982), 238-239.

13. Edward Jay Allen to William Allen, June 23, 1852, Yale Scrapbook.
14. From Fort Laramie to the Mormon Ferry near present-day Casper, Wyoming, Allen's wagon train followed Child's Cutoff on the north bank of the Platte River.
15. The Oregon Territory was formed in 1848. In 1852 it consisted of the current states of Washington, Oregon, Idaho, western Montana, and western Wyoming. South Pass marked the emigrants' entry into the Oregon Territory. As it also marked the Continental Divide and the crossing of the Rocky Mountains, it was a milepost noted in almost every account written by the emigrants.
16. The "Parting of the Ways" was just west of South Pass and marked a split in the trail, the southern route going to Fort Bridger, the northern route known as the Sublette Cutoff going over Dempsey Ridge.
17. The inscriptions are dated July 7. According to Allen's diary, the wagon train departed the morning of July 7, leaving little time to do the inscriptions, which are quite intricate. Probably the majority of the work was actually done on July 6, their rest day. Randy Brown, *Historic Inscriptions on Western Immigrant Trails* (Independence, MO: Oregon-California Trails Association, 2004), 238.
18. Co-author Dennis Larsen, having stood on the top of Dempsey Ridge at the same time of the year, can testify to the accuracy of Allen's account. Edward Jay Allen to William Allen, July 22, 1852, Yale Scrapbook.
19. Allen started at Three Island Crossing. For an overview of the area, see map 88 in Gregory Franzwa, *Maps of the Oregon Trail* (St. Louis: Patrice Press, 1990).
20. Floating down the river with Allen were Martin Koontz, his wife and two young children, Jacob Resser, J. William Carnahan, William McClure, Abe Godfrey, John Watkins, Absalom Dorr and Richard Crist. Larsen and Johnson, 148.
21. Ezra Meeker, *The Busy Life of Eighty-Five Years of Ezra Meeker* (Indianapolis: Wm. B. Burford, 1916), 50.
22. The Dalles refers to the great falls of the Columbia River where it cuts through the Cascade Mountains. The name is derived from the French word "dalle," a stone used to pave gutters. Early travelers saw a resemblance to a gutter in the basalt formations along the narrow river channel. The falls have since been flooded by The Dalles Dam, completed in 1957. Lewis A. McArthur, *Oregon Geographic Names* (Portland: Oregon Historical Society Press, 2003), 945. Celilo Falls, just east of the rapids, was a premier Native American fishing site and has been claimed to be the oldest continuously inhabited dwelling site in North America. William Dietrich, *Northwest Passage: The Great Columbia River* (New York: Simon & Schuster, 1995), 52.
23. A hot, violent wind of the African and Asiatic deserts.
24. Most likely a transcription error. This should probably read "in the winding of the hills."
25. From *Nicholas Nickleby* by Charles Dickens. One of the characters in the novel, Wackford Squeers, operated an abusive all-boys school in London in which the dinner fare was skimpy indeed.
26. Isaac Smith was born in 1825 in Tennessee; his wife Sarah was born in 1830, also in Tennessee. They were married in 1847, and eventually had seven children. Smith was a blacksmith and carriage maker; the family lived in Oregon and California. Stephenie Flora, "Isaac Smith, Pioneer of 1852."
27. The Cascades were a series of rapids just southwest of present-day Stevenson, Washington, created by the debris of a massive landslide which occurred circa 1450 A.D. Very few emigrants attempted to run these rapids. Most disembarked on the north bank of the Columbia River and portaged around them. The rapids disappeared with the construction of Bonneville Dam. McArthur, 177.
28. The famous Hudson's Bay Company post on the north bank of the Columbia River was completed circa 1825, but later moved to higher ground. Now the Fort Vancouver National Historic Site, it includes a re-creation of the original buildings, complete with log palisade—a must-see for anyone interested in regional history.
29. The Willamette is the largest river entirely within the state of Oregon, flowing north from its source through the fertile valley of the same name and entering the Columbia River just north of Portland. Its derivation is from an Indian word, which may mean "spill water." The name was used by David Thompson in 1811, and the currently accepted spelling was used by Charles Wilkes in 1841. Lewis and Clark referred to this river as the Multnomah. McArthur, 1039-1040.
30. A river in West Virginia and southwest Pennsylvania; it joins the Allegheny River at Pittsburgh to form the Ohio River.
31. The *Lot Whitcomb* was a large steamer. Her first route was down the Columbia to Astoria, but soon after her inaugural run, she ran aground, tearing a hole in her hull. After repairs were effected, she ran on the lower Columbia. In 1854, she was sold to the California

Steam Navigation Company, and she ended her days in California running between Sacramento and San Francisco. E. W. Wright, ed., *Lewis and Dryden's Marine History of the Pacific Northwest* (Seattle: Superior Publishing Company, 1967), 29-30.

32. This should read Dr. Ballard, a medical doctor. Dr. David Wesley Ballard settled in Linn County, Oregon. On the recommendation of Oregon Senator George Williams, President Andrew Johnson appointed Ballard as the Territorial Governor of Idaho. Ballard served in that capacity from 1866-1870. After his term as governor, he returned to Oregon. Larsen and Johnson, 179.

Chapter 2: Portland and the Cowlitz Trail

1. A version of this chapter was published in the March 2009 issue of the *Cowlitz Historical Quarterly*, Cowlitz County Historical Society, and is reproduced here with permission.
2. Portland. Established and named in 1845 by Francis Pettygrove. McArthur, 778.
3. Prohibition law. In 1851, Maine passed one of the first statutory implementations of the temperance movement in the U.S. It became known as the "Maine Law." Jim Brunelle, "The Maine Law," www.maine.gov.
4. Although Allen was enthusiastic about the anti-slavery laws in Oregon, the territory had a discriminatory history regarding blacks. In 1844, the Provisional Government of Oregon declared slavery illegal, and settlers who owned slaves were given three years to set their slaves free. All blacks were thereafter banned from residing in Oregon Territory. Black children could stay in the Territory until they reached the age of 18. A variety of exclusion laws followed. In 1849, the legislature enacted another law that forbade blacks from moving to Oregon Territory, although any blacks already in residence were permitted to stay. This act was repealed in 1854. In 1857 Oregon ratified its state constitution, which rejected slavery but retained an exclusion law. Thus, when Oregon became a state in 1859, it had the dubious distinction of being the only state ever admitted to the union with an exclusion law in its constitution. Although that law was superseded by later federal statutes, the wording was not officially removed from Oregon's Constitution until 1927. Elizabeth McLagan, "The Black Laws of Oregon: 1844-1857," BlackPast.org.
5. Term for a two-decked sailing ship, which carried 74 guns. "Boat and Ship Types." www.abc.se.
6. Edward Jay Allen to William Allen, October 15, 1852, Yale Scrapbook.
7. "Territory of Columbia." *Columbian,* April 9, 1853, 2.
8. John Pollard Gaines (1795-1857), the second Governor of Oregon Territory, served from 1850 to 1853. He was appointed to the post by President Zachary Taylor, only after a lawyer by the name of Abraham Lincoln was offered the governorship and declined. John Calvin Gaines III, "John Pollard Gaines," www.salemhistory.net.
9. Joseph Lane (1801-1881) was the first Governor of Oregon Territory. Lane was appointed by President James Polk, and served as Governor from 1849 to 1850, when he became the Territory's delegate to Congress. Lane became one of the two senators from Oregon State in 1859. Fred Blue, "Lane, Joseph (1801-1881)," www.oregonencylopedia.org.
10. Thomas Corwin (1794-1865) served as a prosecuting attorney; a member of the Ohio House of Representatives, United States House of Representatives, and United States Senate; Governor of Ohio; and Secretary of the Treasury. "Thomas Corwin," ohiohistorycentral.org.
11. Allen's manuscript version of his Oregon Trail letters identifies this man as Colonel Sam Black. Pittsburgh Manuscript, 200. Samuel Watson Black was born in Pittsburgh in 1816. He became an attorney, and in 1852 ran unsuccessfully for the U.S. House of Representatives. He later became the seventh governor of Nebraska Territory. He was killed in 1862 during the Civil War battle at Gaines' Mill. "Black, Samuel Watson," politicalgraveyard.com.
12. On foot.
13. From "Don Juan" by Lord Byron.
14. From "The Sexton's Daughter" by Archaeus, *Blackwood's Magazine, Volume 44*, 1838.
15. Early newspapers regularly published alphabetical lists of people who had letters waiting for them at a local post office. "Allen, Edw. J." appeared in the *Oregon Statesman* (Oregon City) lists, beginning on August 15 and ending on September 30. This coincides nicely with Allen's reference to picking up mail from Oregon City. Before he set out on his trip west, he had advised his family and friends to direct mail to the Oregon City post office.
16. Pittsburgh Manuscript, 201-203.
17. Steilacoom is the oldest incorporated town in Washington, and still exists on Nisqually Reach in southwest Pierce County. Its name was taken from a local Indian tribe, and was variously spelled Chil-a-coom, Chel-a-cum, Tail-a-coom, Skil-a-coom, Stiil-le-quem, and Ch-til-acum, in addition to the various spellings Allen threw into the mix. At any rate, the native name means "flowers here." Robert Hitchman,

Place Names of Washington (Tacoma: Washington State Historical Society, 1985), 288-289.

18. Oregon City was situated on the falls of the Willamette River, which provided hydropower for several mills. The town was laid out and named in 1842 by John McLoughlin, chief factor of the Hudson's Bay Company's Fort Vancouver. Oregon City served as the first capital city of Oregon Territory; later, the capital was removed to Salem. The first post office was established in 1847. McArthur, 729.
19. Wilson's first name was never mentioned in Allen's letters. However, this may have been Enoch H. Wilson, who arrived in Oregon Territory in October 1852. Wilson took a donation land claim in the Tumwater/Olympia area; his wife, Anna, died within a year of their arrival in the Northwest. *Washington Territory Donation Land Claims* (Seattle: Seattle Genealogical Society, 1990), 17.
20. Warbassport was a tiny pioneer settlement located on land owned by Edward Warbass, about one and one-half miles downstream from present-day Toledo, Lewis County. Warbassport was next door to another settlement known as Cowlitz Landing; the two were often jointly referred to merely as Cowlitz Landing. Karen L. Johnson, "Forts Along the Cowlitz," *Cowlitz Historical Quarterly, Vol. 1, No. 1* (Kelso, WA: Cowlitz County Historical Society, June 2010), 42.
21. This should be Saturday, and indeed in Allen's later manuscript, he made that correction. Pittsburgh Manuscript, 205.
22. The "young fellow" was Isaac Smith. See Chapter 1, note 26.
23. These sketches were in the possession of Winthrop Allen (Edward's grandson) in the 1950s. He made them available to Blanche Billings Mahlberg for use in her articles about Edward. The authors located photographic prints of a few of the sketches that were, in 2010, in the possession of Blanche's granddaughter. However, the Fort Vancouver sketch was not among them.
24. This should be chief factor. Allen made this correction in his Pittsburgh Manuscript, 206.
25. New Caledonia was an early name for the area encompassing today's British Columbia. Barry M. Gough, "New Caledonia." The Canadian Encyclopedia. http://thecanadianencyclopedia.com/articles/new-caledonia.
26. In his letter and in his Pittsburgh Manuscript, Allen continued with a lengthy tirade on the actions of the Hudson's Bay Company.
27. Nancy Wood Hale and her husband Moses traveled in Allen's Oregon Trail wagon train. Allen gave them use of his oxen when he decided to turn his wagon into a raft and float down the Snake River from Three Island Crossing to Fort Boise. Nancy's husband died of cholera just east of Fort Boise. Allen reclaimed his oxen at the Columbia River Cascades. Larsen and Johnson, 46, 147, 164, 225.
28. Martin Koontz. Allen and the Koontz family (Martin, Mary Ann Kitterman Koontz and their two small children) had journeyed down the Snake (Lewis) River together in wagon-bed boats. Martin and his family later took a donation land claim near Harrisburg, Oregon. The family then removed to The Dalles, Oregon and finally settled in Santa Fe Springs, California. The "young man," whom Allen does not identify, drove Allen's oxen from the Columbia River Cascades to Fort Vancouver. Larsen and Johnson, 242.
29. A Latin phrase translated as "Those who hurry across the sea may change the sky, but not their souls" or "You must change your disposition, not your sky."
30. Pittsburgh Manuscript, 214.
31. Ibid.
32. The North Fork of the Lewis River begins on the slopes of Mt. Adams, and empties into the Columbia River at Woodland. The North Fork was also known as the Cathlapootle River. The East Fork flows through Clark County and joins the North Fork about three and one-half miles above its mouth. Hitchman, 160-161.
33. Pittsburgh Manuscript, 216.
34. The Kalama River was occasionally referred to as the Pretty Girl River. According to Hitchman, 141, "The name is from the Indian word *Calama*, meaning pretty maiden." In *They Came to Six Rivers*, 94, author Virginia Urrutia stated, ". . .Gabriel Franchere in 1811 wrote of the Indian village at the mouth of the Kalama and he said it was called Thlakalamah." However, other sources claim the river and town were named for John Kalama, a Kanaka (Hawaiian) who came to the area circa 1830.
35. In the Pittsburgh Manuscript, 217, Allen corrected this to "a Frenchman, an old Company man."
36. This Frenchman was likely Antoine Gobar (also known as Anton Gobin or Gobain). Gobar was a stockman for the Hudson's Bay Company, and a farmer who lived for several years near the mouth of the Coweman River. So identified was he with this river that locals long called it Gobar's River. Gobar was married to a native woman named Angeline; they had at least five or six children at the time Allen visited them. The family later moved north to Lewis County. Virginia Urrutia, *They Came to Six Rivers: The Story of Cowlitz County* (Kelso, WA: Cowlitz

County Historical Society, 1998), 29, 39; Washington Territory and U.S. Censuses, 1857, 1860, 1870.

37. Probably the snowberry, *Symphoricarpos albus*, a deciduous shrub bearing small flowers, and white berries which hang on late into winter. C. Leo Hitchcock and Arthur Cronquist, *Flora of the Pacific Northwest* (Seattle: University of Washington Press, 1973), 453.
38. Pittsburgh Manuscript, 218-219.
39. Monticello is a very early town near the mouth of the Cowlitz River, established by Harry Darby Huntington. It was twice washed away by floods on the Cowlitz River. Today the city of Longview and course changes in the Cowlitz River have obliterated the original townsite. Hitchman, 192. Allen's recollection of the Hudson's Bay Company sheep drive is confirmed by entries in the *Fort Nisqually Journal.* George Dickey, ed. *The Journal of Fort Nisqually Commencing May 30, 1833, Ending September 27, 1859, Section 9* (Tacoma: Fort Nisqually Historic Site, Metropolitan Park District of Tacoma, 1989).
40. Allen meant, in the conventionalized spelling of the Chinook jargon, "Potlatch hi-yu muck-a-muck" which translates as "Give plenty food." Indeed, he corrected the phrase in his later manuscript. Pittsburgh Manuscript, 221.
41. The "fork of the Cowlits" was the confluence of the Cowlitz and Toutle Rivers. Here lived James "Hardbread" Gardiner, who ran something close to a hotel for travelers, many of whom remarked on the inedibility of his bread. Gardiner later moved to Olympia and died by his own hand in 1858. Gardiner was only forty-five years old at the time of Allen's visit; perhaps his personal habits (if they were on a par with his cooking skills) made him seem an old man compared to Allen's twenty-two years. Karen L. Johnson, "James Gardiner," *Cowlitz Historical Quarterly, Volume 52, Number 2* (Kelso, WA: Cowlitz County Historical Society, June 2010), 4-39.
42. Allen did not consistently capitalize the word Indian. His original spellings are retained here.
43. Pittsburgh Manuscript, 221-222.
44. John McPhail joined the Hudson's Bay Company as a laborer in 1832 and spent most of his years as a shepherd at Fort Vancouver, Cowlitz Farms, and Fort Nisqually. In 1846, he married Therese Cascade. In the 1850 Lewis County, Oregon, census, he was listed as a shepherd. McPhail retired to Vancouver Island in 1853, but later returned to Washington Territory. He died December 15, 1876, in Pierce County, having broken his neck in a fall from a load of hay. "On Thursday . . .," *Puget Sound Weekly Courier*, 4; Harriet Duncan Munnick, *Catholic Church Records of the Pacific Northwest, Vancouver and Stellamaris Mission* (St. Paul, Oregon: French Prairie Press, 1972), A-55-A-56.
45. Pittsburgh Manuscript, 223.
46. The correct spelling is Kanakas, meaning people from Hawaii, then known as the Sandwich Islands.
47. Eugene L. Finch, who also attended the Monticello Convention, was born circa 1819 in New York. He lived in Lewis County for some years, served with the volunteer military during the Indian uprising, and was listed in censuses as a farmer, lumberman, and speculator. *Washington Territory Donation Land Claims*, 148.
48. Samuel Augustus Mitchell, Senior and Junior, were mapmakers and publishers in Philadelphia. Their well-known "New Universal Atlas" was updated annually, and served as a standard among travelers. "Samuel Augustus Mitchell." http://www.raremaps.com.
49. Cowlitz Farm was the vast acreage farmed by the Hudson's Bay Company. The farm produced sheep and all manner of crops, much of which was sold to the Russians in Alaska. Johnson, "Forts," 42.
50. This was probably the tavern known as Goodell's or E. D. Warbass's. Another public house known as Clark's was located on the neighboring land claim. Johnson, "Forts," 44.
51. Quincy A. Brooks was born circa 1826 in Washington County, Pennsylvania, and came west in 1851. He took a donation land claim near Olympia. He was an attorney, customs inspector and deputy customs collector, prosecuting attorney, and special agent for the postal service. In 1857, he moved to Salem, Oregon and in following years resided in several Northwest towns. By 1889, he was collector of customs at Port Townsend. "Pioneer Brooks Dies at Port Townsend," *Morning Olympian*, July 8, 1908, 1; Alan H. Patera, *Your Obedient Servant* (Lake Oswego, OR: Raven Press, 1986), 1-4.
52. Washington, DC.
53. From "Twelfth Night" by Shakespeare.
54. Paraphrase of "An Essay on Man" by Alexander Pope.
55. In his Pittsburgh Manuscript, Allen added, "Smoking was objectionable to no one; possibly some of the more provident had a flask or two that was not kept selfishly for their own use. There was no disposition to drop into sudden slumber, if any felt so inclined he was reminded it was not in good form, and jest and song and story filled up the genial hours." Pittsburgh Manuscript, 230.
56. Probably Alexander William Doniphan (1808-1887), a brigadier general in the Missouri State Militia in the late 1830s, and a colonel in the Mexican-American War.

57. "Ayant the twall" refers to the hours "near or against the twelve" (midnight).
58. The words to the nostalgic song "Ben Bolt" were written by Thomas Dunn English in 1842. They were set to the music of a German melody, adapted by Nelson Kneass.
59. A small river in western Pennsylvania.
60. Having canoed down to Monticello with the group of delegates, Allen returned upriver with some of them at the conclusion of the convention. Since Allen was by this time a sturdy fellow, he must have been hired to help row the large canoe against the strong current of the Cowlitz River.
61. Probably the Douglas fir, *Pseudotsuga menziesii*. Early settlers and loggers often called it the "Oregon pine." Hitchcock and Cronquist, 63.
62. Bran mixed with coarse meal, germ, and/or flour.
63. Mount St. Helens, an active volcano in the Cascade Mountains.
64. John R. Jackson was born in 1800 in England, and came to the U.S. in 1833. He lived for a time in Illinois, but headed west to Oregon Territory in 1844. He established a land claim several miles north of Warbassport. He and his wife, Matilda Glover Koontz (as far as can be determined, no relation to the Koontz family Allen knew on the Trail) were well-known for their hospitality and fine food. Today, the "Jackson courthouse," built circa 1850 (at least one other cabin predated the extant courthouse), is the last remnant of the once-extensive Jackson farm, and is one of the oldest buildings in Washington (although little of the original building materials remain after extensive refurbishing in the early 20th Century). Serene A. Johnson and Karen L. Johnson, "Louisa Jackson's Diary of 1865," *The Lewis County Historian, Vol. 26, No. 2* (Chehalis, WA: Lewis County Historical Society, May 2004), 1.
65. The cat was purchased from James "Hardbread" Gardiner. Pittsburgh Manuscript, 233.
66. The Newaukum is a small river lying entirely within Lewis County, and emptying into the Chehalis River just southwest of the city of Chehalis. The name, often misspelled or mispronounced, means "gently flowing water." Hitchman, 207.
67. Pittsburgh Manuscript, 235.
68. The Skookumchuck River empties into the Chehalis River in present-day Centralia. "Skookum" means strong, and "chuck" means water. The "awful" woods and prairies crossed by Allen before reaching the Skookumchuck would have been the legendarily wet swamplands around present-day Chehalis and Centralia. Hitchman, 277.
69. Joseph and Mary Borst, like the Jacksons farther south, had a reputation for open hospitality and good food. Borst was born in New York in 1821, and came west in 1845. He met and married Mary Adeline Roundtree in 1854. Their home was located on the Cowlitz Trail/Military Road through southwest Washington. Herndon Smith, *Centralia the First Fify Years* (Centralia, WA: The Daily Chronicle and F. H. Cole Printing Company, 1942), 100-135. Although the first home of the Borsts has long since disappeared, their later home built circa 1860 still stands in today's Borst Park in Centralia.
70. The Chehalis River—according to *Place Names of Washington*, "an actual total of 32 spellings of Chehalis are on record." Chehalis means "shifting sands" or "shining sands." The river's headwaters are in the Willapa Hills in southwest Washington; the river flows north into Thurston County, then west into Grays Harbor. Hitchman, 44.
71. Sidney Ford Sr. came to Lewis County in 1845. He was described as "handsome, above middle height, fair and florid of complexion, broad-shouldered, and big around the chest." He established a land claim north of present-day Centralia. During the Indian uprising of 1855-56, Ford showed a deep respect toward the local Native Americans, which they returned in kind. He is credited with helping to quell any hostilities in the vicinity. Smith, *Centralia*, 73-99.
72. Paraphrase of "The Story of Rimini, Canto III" by Leigh Hunt.
73. Allen referred to the mounds today known as the Mima Mounds, a geologic formation that covers many acres in southern Thurston County. Amazingly, at his first sight of the mounds, Allen managed to set forth two theories of formation that are still being endorsed. Varying formation theories include ancient fish nests, gopher mounds, Indian burial mounds, glacial outwash deposits, and polygons created by alternate freezing and thawing. "Mounds of Mystery."
74. Having a piercing glance.
75. New Market was the original name of present-day Tumwater, sited at the foot of the Deschutes River falls. Hitchman, 310.
76. The southern tip of Puget Sound was named for Thomas A. Budd, acting master of the *Peacock*, one of the ships of the Wilkes Expedition of 1841. Hitchman, 30.
77. The "Shute" River is now known as the Deschutes ("waterfall" or "cataract"), which empties into Puget Sound at Tumwater. The falls of the Deschutes enticed pioneer Michael T. Simmons to settle here and build a water-powered mill. Hitchman, 70.

78. The Olympia Hotel (merely a rustic cabin at this time) was built and owned by Edmund Sylvester, credited with founding the city of Olympia. George E. Blankenship, "Lights and Shades of Pioneer Life, By a Native Son," olympiahistory.org.
79. In his manuscript, Allen added, "You must remember that these emigrants left a territory in the east because they deemed the best opportunities were gone, while a European would have thought the ground was scarcely broken, and they did not come here for second choice." Pittsburgh Manuscript, 239.
80. This projected road would eventually run from Steilacoom on Puget Sound, over the Naches Pass in the Cascade Mountains, to Walla Walla in eastern Washington.
81. Isaac Ingalls Stevens, first governor of Washington Territory.
82. More commonly known in the Northwest as the madrona, *Arbutus menziesii.* Hitchcock and Cronquist, 341-342.
83. William H. Aspinwall (1807-1875) was born in New York, and became a shipping magnate. In 1848 he founded the Pacific Mail Steamship Company to carry mail between Panama and the newly-annexed territory of California and Oregon Territory. "William Henry Aspinwall."
84. The second largest island in the lower 48 States, Whidbey Island was named by Captain George Vancouver for Joseph Whidbey, master of the ship *Discovery.* Hitchman, 329. There is a Whitney Island in Padilla Bay off Bellingham, but this is a very small island; Allen was obviously mistaken about the name, or the *Pittsburg Dispatch* typesetter misspelled it. In his Pittsburgh Manuscript, Allen corrected the name to Whidbey. Pittsburgh Manuscript, 241.
85. The authors were unable to locate any present-day reference to this phenomenon. It may, however, have been an underground fire in a coal seam.
86. Pittsburgh Manuscript, 242.
87. Ibid.
88. Ibid.
89. The thimbleberry, *Rubus parviflorus*, salmonberry, *Rubus spectabilis*, and salal, *Gaultheria shallon*, are all fruit-bearing species, important for people and wildlife. Hitchcock and Cronquist, 224, 343-344. In a letter from 1853, Allen's sister Rebecca wrote of receiving some seeds from him: "The seeds you sent were planted, but not one germinated. When they reached us they were all in one mouldy solid cube. Mother thinks very likely the paste with which you pasted them up may have dampened them and caused them to mold. We should have been so much pleased if they had grown, if only for your sake." Rebecca Allen to Edward Jay Allen, September 27, 1853, Rosanio Collection.
90. This may be *Petasites frigidus*, sweet coltsfoot. Other possibilities include *Achillea millefolium*, common yarrow; *Tanacetum douglasii*, western tansy; or *Achlys triphylla*, vanillaleaf; however, none of these meet all of the adjectives applied by Allen to his specimen. Scott Clay-Poole, Ph.D., has written an ethnobotany essay for the Washington State Department of Transportation website. In a 2013 interview, he stated that sweet coltsfoot best fits Allen's description.
91. Allen may have referred to the celery-leaved licorice-root, *Ligusticum apiifolium.* Hitchcock and Cronquist, 327. Other plants also known as licorice-root would not be native to a seashore environment.
92. From *The Use of Flowers* by Sarah Josepha Buell Hale.
93. Lemingia is an unknown reference. *Ranunculus* is the buttercup. Trillium is a familiar northwest wildflower, and lupines are abundant in prairie and alpine habitats. Hitchcock and Cronquist, 124, 265, 695.
94. In 1840, Alvin Adams began a fairly local express business out of Boston. By 1850, he had expanded his delivery lines as far south as St. Louis. A subsidiary concern was organized in California in 1850 and was soon servicing towns up and down the Pacific coast. "The Adams Express Company." Well-known Olympia resident Captain John G. Parker, however, stated that he came west in 1851 as a messenger for Gregory's Express, the first express company established on the Pacific coast. By late spring of 1853, Parker had advanced to the Olympia area and established a regular express "connecting with Adams & Co.'s" at Portland. Robert A. Bennett, ed., *A Small World of Our Own*, (Walla, Walla, WA: Pioneer Press Books, 1985), 64-69. Adams & Co. was advertising its newly-established Portland office in Olympia's *Columbian* as early as September 18, 1852.
95. In 1851, the sloop *Georgiana* sailed north from Olympia on a gold-seeking expedition. Upon reaching the Queen Charlotte Islands off the coast of British Columbia, the ship was driven ashore by a storm, and the crew members were taken prisoners by the Haida Indians. Bennett, 97-100. A rescue party succeeded in ransoming the crew, one of whom was Charlie Weed, Allen's partner in exploring Puget Sound (see Chapter 7). Allen's Pittsburgh Manuscript gave an amusing description of Weed's experiences.
96. William Billings was born in 1827 in Ripton, Vermont, arrived in Portland in 1849, and came to Thurston County in 1851. He held several public offices, most famously, Thurston County Sheriff, a position he held

for twenty-four years. He also partnered with two other area residents to establish the Territory's first penitentiary at Seatco (now Bucoda), several miles south of Olympia. Billings died in 1909. "Thurston County Pioneers Before 1870," Washington Rural Heritage, washingtonruralheritage.org.

97. Argillite, which is found only on the Queen Charlotte Islands; the supply of argillite is strictly controlled by the Haida Indians. The Haidas carved very complex figures and pipes from argillite, and sold them to crews of visiting ships. The carvings seemed to have been intended primarily as trade items, rather than ceremonial objects for use within the tribe. "Haida & Argillite," Simon Fraser University, www.sfu.ca.
98. Madoc was a Welsh prince who, according to legend, discovered America circa 1170. He and his crew have been credited with establishing colonies in various areas of North America, and with spreading the Welsh language to several Native American tribes. In Allen's time, this legend was on the way to being wholly discredited. David B. Quinn, "Madoc," Dictionary of Canadian Biography Online.
99. Pittsburgh Manuscript, 244, 245.
100. A river in western Pennsylvania, which joins the Monongahela River at Pittsburgh to form the Ohio River.
101. A packet steamer built in 1848 at Elizabeth, Pennsylvania. "Steamboat Building in Elizabeth, PA," *Built at Elizabeth PA.*
102. Washington Territory's first newspaper, established in 1852 in Olympia by Thornton McElroy and J. W. Wiley.
103. From "The Little Castle Builders" by Thomas Bailey Aldrich. Pittsburgh Manuscript, 247.
104. From "Childe Harold's Pilgrimage" by Lord Byron.

Chapter 3: At Home in Olympia

1. "Below," in this context, refers to a location that is lower than the head of a body of water. The head is the highest point, which in Puget Sound's case is the southernmost tip at Olympia. Allen's claim, which was three miles *north* of Olympia, is in a maritime sense *below* Olympia. Thus, everything south of Allen's land was "above" his claim, and everything north was "below" his claim.
2. Archaic word meaning "to put in writing."
3. Eld Inlet is one of the fingers of water making up southern Puget Sound. It was named for Midshipman Henry Eld of the Wilkes Expedition of 1841. Hitchman, 81.
4. The four years' residency requirement was later decreased to one year and the payment of $1.25 per acre. Allen took advantage of this change; he was able to leave his claim in early 1855, having complied with the amended conditions. His land patent was finally issued in 1869. *Washington Territory Donation Land Claims*, 1.
5. Shirley Ensign was born circa 1828 in Ohio, and was listed as a farmer and a miner in various Washington Territory censuses. He also built and operated a ferry on the Columbia River in 1853-54 under Allen's charge from the Territorial Legislature. In 1855, he was appointed sheriff for Walla Walla County. He served in Company C, Mounted Rangers, during the 1856 Indian uprising. "Military Department, Indian War Muster Rolls, 1855-1856 – Shirley Ensign," Washington State Archives, Digital Archives; Washington Territory censuses.
6. Allen identified the Dutchman only as Fred from Beirenth, Bavaria. Attempts to further identify Fred have been unsuccessful.
7. "Skookum," Chinook jargon for strong or good.
8. The Free Soil Party was a relatively short-lived political party, most active in the 1848 and 1852 presidential elections. It opposed slavery, particularly its expansion into the western states and territories. Its slogan was "Free Soil, Free Speech, Free Labor and Free Men." It was largely absorbed by the Republican Party in 1854. "Free Soil Party," *Encyclopedia Britannica.*
9. Several paragraphs later, Allen identifies this fellow as "Mumpford" who remains a rather cryptic person. A William Mumford appeared in Thurston County court records in 1854, but it is not certain if this is the same man of Allen's acquaintance.
10. "Muck-a-muck" meant "food" in the Chinook jargon.
11. Michael Troutman Simmons is widely credited with establishing the first American settlement (New Market, now Tumwater) north of the Columbia River. Simmons was born in Kentucky, and arrived on Puget Sound in 1845. The next year he established a saw and grist mill at the falls of the Deschutes River. He was a delegate to the Monticello Convention, and later removed to Mason County. He spent his last few years in Lewis County, where he died in 1867. Gordon Newell, "So Fair A Dwelling Place," Olympia Historical Society, olympiahistory.org.
12. "Mrs. Partington" was a character created by the humorist Benjamin P. Shillaber (1814-1890) who wrote for the *Boston Globe* in the 1840s. Mrs. Partington was a narrow-minded literalist and user of malapropisms—words that sound like the one intended but that are comically misused. The columns of the *Globe* were filled with her witty, comic monologues. "Benjamin Penhallow Shillaber," Chelsea, Massachusetts, Historical Society.

13. From "Friends" by Francis Brown.
14. From "The Norseman" by John Greenleaf Whittier, *Knickerbocker, Volume 17*, 1841, 16.
15. From "Sabbath Day," *London Journal.*
16. Pittsburgh Manuscript, 252.
17. Chinook jargon for "dead" or "kill."
18. Pittsburgh Manuscript, 253.
19. In his later manuscript, Allen wrote: "The house when finished was primitive enough, being a balloon frame, of rough boards on end, and one story, with a little stoop in front, and a shingled roof, with a steep pitch, and no battens except in the corner round the stove, because the wind flickered the candles too much, the other openings were not battened and with the unglassed window insured a good ventilation day and night." Pittsburgh Manuscript, 253.
20. Allen did not identify this young fellow.
21. Dr. Albert G. C. W. Eggers was born in Germany, lived in Pittsburg, and came west over the Oregon Trail in 1851. After arriving in the Olympia area, he took up a small land claim, and applied for a land patent in September 1869. During the Indian uprising of 1855-56, he was an assistant surgeon in the local military. He was also a musician, orchardist and medical doctor. *Washington Territory Donation Land Claims;* "Military Department, Indian War Muster Rolls, 1855-1856—Albert Eggers," Washington Territorial censuses.
22. Eggers and Brooks did travel west together. On April 2, 1851, the *Pittsburgh Gazette and Advertiser* wrote "Albert Eggers M.D. and Quincy A. Brooks, Esquire, attorney and counselor at law, two of our most estimable citizens, leave today, on the Oriental [a ship], *en route* for Oregon. We trust that they may there meet with that success which their merits so well deserve." Eggers and Brooks traveled by wagon train over the Oregon Trail, and arrived in the Portland area in late 1851. "For Oregon," *Pittsburgh Daily Gazette and Advertiser*, April 2, 1851, 3.
23. Allen did indeed purchase land for his brother William, although not an entire land claim; various parcels were purchased in areas that today comprise downtown Olympia. *Thurston County Deeds, Vol. 1,* Washington State Archives, Olympia, 54-55, 86-87, 175-176, 182-183.
24. From "The Brothers" by Charles Sprague. Allen's choice of poetry was particularly apt here. He referred to the death of his brother, George, who died in a steamboat accident in 1847, leaving Edward and William as the two brothers in the family.
25. Allen did indeed ask his brother William to investigate and then invest in a sawmill. William viewed model sawmills at the Pennsylvania State Fair, and also visited operating sawmills around the Pittsburgh area. Despite the avid investigations, the project never materialized. Rebecca Allen to Edward Jay Allen, September 27, 1853, and April 15, 1854, Rosanio Collection.
26. Now known as Tykle Cove. George Tykle (also spelled Tykel) was born in Germany, traveled to the Northwest, became a naturalized U.S. citizen in 1853, and took a donation land claim next to Allen's. *Washington Territory Donation Land Claims*, 98.
27. Mount Rainier. In the Pittsburgh Manuscript, 258, Allen typed "Rainier," then crossed it out and wrote "Tacoma" instead. One of the early native names for the mountain was Tacoma or Tahoma, meaning "snow-covered mountain." In 1917, the U.S. Board of Geographic Names approved Rainier as the official name, which was originally applied by Captain George Vancouver in 1792, to honor Rear-Admiral Peter Rainier. Hitchman, 198. Since then, many Washingtonians have lobbied to change the name to Tahoma or Tacoma.
28. Archaic word meaning reward or merit.
29. Matinal, meaning of the early hours (breakfast time).
30. Pittsburgh Manuscript, 256-259.
31. Chinook jargon for "no good."
32. Pittsburgh Manuscript, 261-262. In his work "An Essay on Man: Epistle I," published between 1732 and 1734, Alexander Pope wrote "Lo! The poor Indian, whose untutor'd mind…," characterizing Indians as having simple natures with few desires or ambitions. The phrase caught on, especially in America. Horace Greeley utilized it as the title of one of his letters from the West serialized in the *New York Tribune* in 1859. Many other books, essays, and plays made use of the phrase. Throughout the last half of the nineteenth century the term "Lo" was used ubiquitously to refer to Indians individually and generally.
33. In his manuscript, Allen explained that the Indian legend actually had a basis in geologic fact. Pittsburgh Manuscript, 262-263.
34. Probably Hewitt Lake, located in a "kettle" depression formed during the glaciated period of Puget Sound's geologic history. The lake has been called "Haunted Lake" as well as Lowe Lake. Gayle Palmer and Shanna Stevenson, eds., *Thurston County Place Names* (Olympia, WA: Thurston County Historic Commission, 1992), 35.
35. Pittsburgh Manuscript, 262-264.
36. From "Hymn of the Cherokee Indian" by I. McLellan.
37. From "Paradise Lost" by John Milton.
38. From "The Disinterred Warrior" by John Elliot.
39. Allen and the Pittsburgh typesetters may have meant "potentates."

40. Sealth, or Seattle, was born circa 1786, and died in 1866. Northwest tribes did not have hereditary chiefs, so he would be more properly considered a leader of the Suquamish and Duwamish tribes. "See-atch" is an anglicized pronunciation of his name. Seattle was very friendly toward the white settlers, who honored his friendship by naming a small village after him. David M. Buerge, "Chief Seattle and Chief Joseph: From Indians to Icons," University of Washington Libraries—Digital Collections, http://content.lib.washington.edu/aipnw/buerge2.html.
41. "Tomanawos" is Chinook jargon for "medicine man" or "magic."
42. From "The Rime of the Ancient Mariner" by Samuel Taylor Coleridge.
43. Allen's extrapolation became an innocent exaggeration. Studies of old-growth trees, growing under the best conditions, show that the tallest native species in western Washington rarely attained 300 feet in height, no matter their base diameter. The Douglas fir, *Pseudotsuga menziesii*, may grow to 325 feet; grand fir, *Abies grandis*, 250 feet; western redcedar, *Thuja plicata*, 250 feet; and western hemlock, *Tsuga heterophylla*, 250 feet. These would be extreme examples of their species; the average for any species would be about 50 feet less than the maximum stated above. W. M. Harlow and E. S. Harrar, *Textbook of Dendrology, 4th Edition* (New York: McGraw-Hill, 1958), various.
44. Pittsburgh Manuscript, 268-269.
45. American spiritualism had its beginnings in a farmhouse in Hydesville, New York, in 1848. The Fox family, particularly two daughters, claimed their house was visited by the spirit of a murdered man, who communicated with the daughters by a series of raps. The alleged events were widely publicized, and spiritualism was born. The movement had many adherents and just as many detractors, including churches which associated the practice with witchcraft. "Spiritualism," *Encyclopedia Britannica*.
46. "Wake" means "no" in Chinook jargon.
47. "Killapie" is Chinook jargon for "overturn."
48. Baron Karl von Reichenbach was a German scientist who conducted many experiments into a mysterious force which he believed permeated nature and could be identified by "sensitives." His research was published and translated into English in 1850; this was probably the work with which Allen was familiar. "Baron Karl von Reichenbach," *Gale Encyclopedia of Occultism & Parapsychology*. Emanuel Swedenborg (1688-1772) was a Swedish scientist, philosopher and theologian who in 1745 claimed to have received direct insight into the spiritual world; following this revelation, he propounded mysticism and spiritualism. "Emanuel Swedenborg," *Gale Encyclopedia of Occultism & Parapsychology*.
49. From "Spiritual Presence" by James H. Perkins.
50. Unknown source.
51. From "The Dark Valley," *Knickerbocker, Volume 39*, 1852, 459.
52. No doubt Allen meant feline.
53. Allen's older brother, William, who was serving as a captain or purser on steamboats.
54. The encampment of Jason Lee in the Blue Mountains near present-day Meacham, Oregon.
55. Pittsburgh Manuscript, 274.
56. Allen's term "fish crow" referred to the bird he would have known from the southeastern U.S., but which does not reside in Washington. What he really observed was either the American Crow, *Corvus brachyrhynchos*, or the Northwestern Crow, *Corvus caurinus*. The call of the Northwestern Crow is similar to that of the Fish Crow. Brian H. Bell and Gregory Kennedy, *Birds of Washington State* (Auburn, WA: Lone Pine Publishing International Inc., 2006), 261, 365; Ted Floyd, *Smithsonian Field Guide to the Birds of North America* (New York: Harper Collins, 2008), 318-320.
57. Pittsburgh Manuscript, 275.
58. Pittsburgh Manuscript, 276.
59. Victoria was, and still is, the principal city of Vancouver Island, British Columbia.
60. See note 95, Chapter 2.
61. "Hi-yu" is Chinook jargon for "plenty."
62. Epicurean; having a refined and discerning taste for fine foods.
63. Pittsburgh Manuscript, 277.
64. In a letter dated March 8, 1853, William wrote to Rebecca, "From Eddie's directing his letters to be directed to Olympia Oregon Territory I am afraid he has not received all we have sent; which according to his instructions were always to be directed to Oregon City till countermanded, don't know how often the mail goes or what distance there is between Oregon City or Olympia, but he'll have to send word to the Post Master there to forward them to him and in all your letters remind him we *have* written often and long, and in magazines and papers have not been lacking, there must be a Bushell at least awaiting him there that I sent him." William Allen to Rebecca Allen, March 8, 1853, Turner Collection, UALR.
65. Pittsburgh Manuscript, 278-279.
66. From "The Land of Gold," R.H. Stoddard, *Knickerbocker, Volume 33*, 1849, 395.
67. "Si-wash" (from the French *sauvage*) was Chinook jargon for an Indian.

68. Paraphrase of "The Veiled Prophet of Khorassan" by Sir Thomas Moore.
69. Paraphrase of "Benedicite" by John Greenleaf Whittier.
70. Allen wrote this letter directly to L. Heron Foster and Reese C. Fleeson, editors of the *Pittsburg Daily Dispatch*. Yale Scrapbook.
71. "Cacoethes scribendi" is Latin for the "irresistible urge to write."
72. Adam Wylie had a land claim across Budd Inlet from Allen's claim. Wylie was born in Ireland circa 1806, arrived on the Pacific coast in 1849, became a naturalized citizen in 1851, and settled on his land claim in 1852. *Washington Territory Donation Land Claims*, 81.
73. The *National Era* was a weekly journal published in Washington, DC, by Gamaliel Bailey.
74. Reverend Pascal Ricard was born circa 1805, became a Catholic priest, and in 1848 established an Oblate mission on the east shore of Budd Inlet, on the site of what is now Olympia's Priest Point Park. *Washington Territory Donation Land Claims,* 26; Hubert Howe Bancroft, *Washington, Idaho, and Montana: 1845-1889, Volume 31* (San Francisco: History Co., 1890), 10.
75. Jacob Resser, J. W. Carnahan, and William McClure were the three friends that started out on the Oregon Trail with Allen in 1852. All arrived safely in Oregon Territory. Marysville, in central western Oregon, later changed its name to Corvallis. (Corvallis city history notes that in 1853, a tin shop was located in town; this was probably Resser's shop.) Jacksonville, in southwest Oregon, was founded in 1851-52. The Umpqua River Valley is also located in southwest Oregon, and no doubt had many lumber and shingle mills. McArthur, 234, 506, 982.
76. From "A Thing of Beauty" by John Keats.
77. "Clatta-wah" is Chinook jargon for "go."
78. Here is a prime example of how things occasionally went wrong between Allen's penmanship and Pittsburgh typesetting. This phrase should read "deavin dinsome" and in Scottish dialect means "deafeningly noisy." The line comes from William Motherwell's poem titled "Jeanie Morrison" and refers to a very noisy town.
79. From "Hidden Life," *Knickerbocker*, 1850, 300.
80. Paraphrase of "Hadad" by James A. Hillhouse.
81. Pittsburgh Manuscript, 285.
82. Pittsburgh Manuscript, 285-286.
83. Pittsburgh Manuscript, 287.
84. The camas, *Camassia quamash*, is one of the few Northwest plants with both a scientific and common name that honor the Indian language. The bulb of the common camas was an important food source for Northwest Indians. Hitchcock and Cronquist, 688.
85. Pittsburgh Manuscript, 288-289.
86. Pittsburgh Manuscript, 290.
87. "[A]ll children born of persons living and cohabiting together, as man and wife, and all children born out of wedlock whose parents shall intermarry, shall for all purposes be legitimate." "An Act to Regulate Marriages," Laws of Washington Territory, 1854, Washington State Archives, 691.
88. Pittsburgh Manuscript, 291.
89. In his Pittsburgh Manuscript, 291, Allen corrected his jargon to "Hyas klosh mamack muck-a-muck," which does indeed mean an excellent cook.
90. A large lake three miles southwest of Olympia. The 1841 Wilkes Expedition used the natives' name, "Sa-chal." Hitchman, 22.
91. In his later manuscript, Allen added: "and it was a grand experience." Pittsburgh Manuscript, 292.
92. Pittsburgh Manuscript, 293.
93. Ibid.
94. The brig *Cyclops* was a regular trader along the Pacific coast. Wright, 40.
95. Beirenth is a town in Bavaria.
96. In his Pittsburgh Manuscript, Allen corrected this to Fitchtelgebirge. Pittsburgh Manuscript, 297. The correct name is Fichtelgebirge, which is a mountain range in Germany. "Fichtel Hills," *Encyclopedia Britannica.*
97. Benjamin F. Yantis was born in Kentucky in 1807, and came to the Northwest in 1852. His wife died on the Oregon Trail journey. Yantis was a Democrat, Presbyterian, and Mason, and served as a justice of the peace, territorial legislator, Olympia city councilman, and Indian agent at Fort Colvile. He was part owner of the first stage line to run south from Olympia. "Judge B. F. Yantis," Obituary. *Washington Standard*, February 15, 1879.
98. From "Art Above Nature" by Peter Pindar.

Chapter 4: The Road Viewers

1. "Road Over the Cascade Mountains." *Columbian,* April 30, 1853, 2.
2. Michael Simmons' route ran from the Puyallup River through today's Bonney Lake and Connell Prairie, until he gave up the effort in the timber section south of present-day Buckley.
3. "The Road Across the Cascade Mountains." *Columbian,* May 28, 1853, 2.
4. "Thurston County Pioneers Before 1870."
5. George E. Blankenship, "Lights and Shades."
6. The term "road-viewers" referred to parties sent out to determine the suitability of a proposed route for use

as a road. It was in common usage in the 1850s and 1860s. Today they would be recognized as highway department surveyors.

7. "Churches of Washington." "Thurston County Pioneers Before 1870. "Died," *Pioneer and Democrat*, October 30, 1857. Jimmie Jean Cook, *"A particular friend, PENN'S COVE:" A History of the Settlers, Claims, and Buildings of Central Whidbey Island* (Coupeville, WA: Island County Historical Society, 1973), 35-37.
8. Pittsburgh Manuscript, 299.
9. *Knickerbocker*.
10. This was the effort made by Michael Simmons; see note 2.
11. Bancroft, in his *History of Washington*, said the meeting was held on May 14. His source was an article in the May 28, 1853, issue of the *Columbian* that told of the meeting. Bancroft apparently misread the article as it said the meeting was held "pursuant to a call in the *Columbian* of the 14th inst…," not held on the 14th. Allen, who was there, said the meeting was held on May 21. Another article in that very issue of the *Columbian* said the meeting was held on May 21, as did yet another *Columbian* article dated June 4.
12. Allen's June 18 report, published in the *Columbian* on July 2, revealed that the party departed from John Edgar's home (near present-day Yelm) on June 5. Working backward from that date, Wednesday, June 1 would have been the date the party intended to depart from Olympia. It is not clear from the letters if Allen spent two or three nights at Edgar's. Three nights' stay would match with a Wednesday, June 1, exit from Olympia.
13. From "Indian Triumph Song," *Knickerbocker*, January 1853.
14. Edward Jay Allen to William Allen, May 22, 1853, Yale Scrapbook.
15. John Edgar was born in England in 1814. He arrived in Oregon Territory on January 1, 1839, and in 1849 married a Nisqually Indian named Elizabeth (Betsy). Edgar became a U.S. citizen in April 1853. Drew Crooks, "Intermarriage: The Example of John and Betsy Edgar," Leschi, Close Ties; *Washington Territory Donation Land Claims*, 95.
16. Whitfield Kirtley was born in Kentucky. After arriving in Olympia, he served as school district director, member of the 1st Regiment Puget Sound Volunteers, and Judge of Election. By 1860, he had moved to Mason County. *Thurston County First Record Book, 1852-1857*, Washington State Archives, Digital Archives, http://www.digitalarchives.wa.gov; Angie Burt Bowden, *Early Schools of Washington* (Seattle: Lowman and Hanford Company, 1935), 131; Washington Territory and U.S. censuses.
17. George Shazer (also spelled Shaser) was born July 17, 1815. In his early life he was an employee of the American Fur Company working primarily around the Great Lakes. He came west over the Oregon Trail in 1845, and along the way (at Independence Rock) was married to Margaret Packwood. After initially settling south of the Columbia River, the couple moved north in February 1847 to a farm near the mouth of the Nisqually River (about where the headquarters building of the Nisqually National Wildlife Refuge is located today). Shazer died in 1899. "Thurston County Pioneers Before 1870."
18. Fort Snelling, Minnesota.
19. In 1838-39, Frémont was with Joseph Nicollet exploring the lands between the Mississippi River and the Missouri River.
20. Edward Jay Allen to William Allen, July 20, 1853, Yale Scrapbook.
21. Named for David and Andrew Chambers, who arrived in 1845. Hitchman, 44. Their claim was located about a half mile south of the present-day intersection of Yelm Highway and Rainier Road. The location is marked with a historical plaque and a new housing development.
22. Charles Eaton was one of Oregon's earliest pioneers, arriving in Yamhill County in 1843. He settled about twelve miles southeast of Olympia in 1847 on what became known as Eaton Prairie. The following year, he went to California in search of gold. Upon his return in 1849 he found squatters on his claim. He relocated on Tenalquot Prairie (near present-day Waldrick Road) and in 1854 married Chief Leschi's daughter, Kalakala. Eaton fought in both the 1848 Cayuse Indian war and in the 1855-56 Puget Sound Indian war. George Himes, "Pioneer Reminiscences," *Transactions of the 53rd Annual Reunion of the Oregon Pioneer Association* (Portland: Oregon Pioneer Association, July 1925), 13.
23. Both Charles and Nathan Eaton settled in the area. Charles' claim, however, fits Allen's description of "bottom land."
24. This lake probably filled in over the years, as it no longer appears on maps. It was located on what is now Fort Lewis property.
25. Edward Jay Allen to William Allen, July 20, 1853, Yale Scrapbook.
26. In his Pittsburgh Manuscript Allen typed Yelm, which was the Native American name for the prairie. The 1853 typesetters in Pittsburgh obviously had difficulty with local names such as this and incorrectly set it as "Yellow."

27. John Edgar's land claim was just north of present-day Yelm, Thurston County.
28. Captain McClellan met chief Owhi, brother of Kamiakin, on August 29, 1853 when he was returning from his visit to the summit of Naches Pass. McClellan called him the "best-natured Indian yet seen in the country." "Longmire Tells Naches History," *Yakima Herald*, March 12, 1923, 1.
29. The story of that trip may be found in "Visit to Mt. Rainier," *Columbian,* September 18, 1852, 2.
30. This is most likely Crystal Spring, located in the north part of the city of Yelm at the end of Crystal Spring Road, just north of the Centralia Diversion Canal.
31. See Chapter 2, note 39.
32. Edgar was well served in making a will. On November 10, 1855 he was wounded in the Indian Wars; he died on November 21, 1855. Dean Hooper and Roberta B. Longmire, *Yelm Pioneers and Followers, 1850-1950* (Yelm, WA: Yelm Prairie Historical Society, 1999), Appendix A.
33. Dr. William Fraser Tolmie, chief factor at the Hudson's Bay Company installation at Fort Nisqually.
34. Probably Henry Smith who had a donation land claim in Pierce County (near present-day Roy). Smith was born in 1824 in Prussia, and arrived in Oregon Territory in September 1851. *Washington Territory Donation Land Claims:* 120. In *Washington Territorial Volunteer Papers, Indian War Correspondence 1855-57*, Washington State Archives, Smith is noted as "Smith, Henry (Sandy)," a Hudson's Bay Company employee at Muck Station.
35. Allen referred to a series of prairies located along the north edge of Muck Creek. The party traveled through these prairies toward present-day Graham where they came to a view of the Puyallup valley above the site of present-day Orting.
36. From "The Fugitive" by William Henry Burleigh.
37. The Puyallup and Carbon Rivers merge just west of Orting, and a number of creeks in the area are possibilities for Allen's four branches of the Chewah or Chew River as it is called in the Pittsburgh Manuscript. He might also have referred to the Puyallup River, as it was much braided at that time, though he specifically mentioned crossing it the following day.
38. This prairie was likely near the present-day city of Orting where the viewers forded the Puyallup River.
39. From "Lalla Rookh, Vale of Cashmere" by Thomas Moore.
40. This is probably the Carbon River.
41. Allen was climbing Prairie Ridge, the hill just northwest of present-day Orting.
42. Allen was being forced back toward South Prairie Creek. Meeker confirmed this in *Pioneer Reminiscences of Puget Sound* (Seattle: Lowman and Hanford, 1905), 135.
43. Small groves of trees.
44. Today Mt. Rainier is measured at 14,410 feet. Allen was actually over fifty miles from the mountain. Its bulk is such that many visitors have assumed it is much closer than it actually is.
45. This is either South Prairie or the prairies around Buckley, perhaps Connell Prairie.
46. Written by John Greenleaf Whittier for the opening of Pennsylvania Hall.
47. Allen probably referred to the road that Michael T. Simmons tried to cut through Naches Pass in 1850. Bancroft, 63. See note 2, this chapter.
48. The White River heads on the Emmons Glacier on Mount Rainier. In June when the melt is at its peak, the White River is formidable indeed. The ford would have been west of present-day Enumclaw.
49. From *Eliza Cook's Journal*, 1850.
50. The area Edgar spoke of is west of today's Enumclaw and is still mostly prairie.
51. Allen joined his brother William on a steamboat trip down the Ohio and Mississippi rivers from Pittsburgh to New Orleans and back in February 1851. Larsen and Johnson, 26.
52. Probably the Snoqualmie River, which is a tributary of the Snohomish River. The Snoqualmie is about 30 miles north of Buckley, which coincides with Allen's location at this time.
53. The narrow canyon is at the location of today's Mud Mountain Dam. Allen exaggerated when calling it a tremendous mountain.
54. Snake River.
55. From "The Walk" by Friedrich Schiller.
56. Pittsburgh Manuscript, 313.
57. Probably Scatter Creek.
58. Probably just west of Federation Forest State Park.
59. Allen's Pittsburgh Manuscript, 315, said, "T'suck Tiee" or water brave.
60. This peak, LeTete, was named by French-Canadian trappers of the Hudson's Bay Company because its summit was shaped like a human head. In the 1850s it was a well-known landmark on the trail over the Cascades; almost everyone who passed by noted it. Neither "Letebe" nor "LeTete" are shown on current USGS maps. Most early sources placed it just northwest of the junction of the Greenwater and White Rivers.
61. Bear Prairie was a small salal-covered flat within the fork of the Greenwater and White Rivers. It was the first place west of Government Meadow where livestock could find grass. Governor Stevens asked the

second Territorial Legislature for funds to seed the prairie with grass as an aid to emigrants. Ezra Meeker, George Himes and Clarence Bagley camped there during their 1919 exploring trip to Naches Pass. Bear Prairie is about three miles east of today's Greenwater.

62. From "May Flower" by Frederick Goddard Tuckerman.
63. "It was an ascent of a thousand feet. I cannot say how many degrees, but I judge a natural slope of one to three, and yet covered from foot to summit with huge trees, two or three hundred feet high." Pittsburgh Manuscript, 319.
64. Government Meadow, a beautiful prairie about a quarter mile west of the summit.
65. It is called Lamoti Klosh Illehe ("good ground of the mountain") in the Pittsburgh Manuscript, 317.
66. This compass is in the Hervey Allen collection at the University of Pittsburgh.
67. Little Naches River. This camp was probably Timothy Meadows.
68. This may be one of three roots. Balsamroot, *Balsamorhiza sagittata*, was an important food source for Native Americans. C. P. Lyons, *Trees, Shrubs and Flowers to Know in Washington* (Toronto: J. M. Dent and Sons Ltd., 1956), 148; Hitchcock and Cronquist, 495. Western spring beauty, *Montia sibirica*, is found in wet gravelly sub-alpine meadows on the east slopes of the Cascade Mountains. Hitchcock and Cronquist, 108. Camas, *Camassia quamash*, blooms in June at the 2,500 to 3,000 foot altitude. Hitchcock and Cronquist, 688. The latter two could be cooked quickly as Allen described. Scott Clay-Poole, Ph.D., has written an ethnobotany essay for the Washington State Department of Transportation website. In a 2009 interview, he stated that western spring beauty best fits Allen's description.
69. From "The Living Age" by Eliakim Littell.
70. "We camped that night upon the Ahnepash, a mile or two above an Indian fishery." Pittsburgh Manuscript, 324.
71. This was probably Edgar Rock (named for John Edgar) near present-day Cliffdale.
72. "They were Nesqully's, in Owchi's country, and not familiar with anything off the trail." Pittsburgh Manuscript, 324.
73. Slithering.
74. This was likely Jane Edgar, who would have been about eight years old at the time. Jane died of typhoid at about age 15. "Died." *Washington Standard*, February 23, 1861, 3.
75. From "Mazeppa" by Lord Byron.
76. The route left the Naches River at Benton Creek and climbed to the divide overlooking the Wenas Valley.
77. See Chapter 3, note 25.
78. The 1853 smallpox epidemic and its spread into the Yakima Valley is documented in Robert Boyd's *Coming of the Spirit of Pestilence* (Seattle: University of Washington Press, 1999), 160-167.
79. A leap made by a horse where all four feet are momentarily off the ground, the forelegs landing first.
80. William Packwood was Margaret Packwood Shazer's uncle and next-door neighbor.
81. Eventually fourteen children were born to the Shazers.
82. Paraphrase of "Lochinvar" by Sir Walter Scott.
83. B. F. Yantis; see Chapter 3, note 97.

Chapter 5: Robbed!

1. Paraphrase of a variation of a "Limber Jim" minstrel song, which was itself an adaptation of the "Froggie Went A-Courtin" song of the 1800s.
2. Pittsburgh Manuscript, 334.
3. Allen's sister Rebecca wrote, "Just a week ago I wrote you a letter enclosing one from William, also $20 and a heavy gold ring. This I hope you have gotten… " Obviously, the ring did arrive safely, only to be stolen from Allen's cabin several months later. Rebecca Allen to Edward Jay Allen, December 28, 1852, Rosanio Collection.
4. In his Pittsburgh Manuscript, 333, Allen added: "[A]nd other little things, that were my evidence that I had once been in the midst of a fuller civilization, and 'moved in the highest circles'; just what the 'moving' consists of I dunno, but if it is done in these surroundings, of course I did it, and these were the things I had on while I was doing of it."
5. On July 2, 1853, the *Columbian* published the following article under the headline "Robbery":

   > The house of EDWARD J. ALLEN, three miles below this place on the west side of the bay, was broken open by Indians while Mr. A. was absent assisting to view out the road across the mountains. A good many articles were stolen, among which the following will be most easily identified: A heavy gold ring with motto engraved on the inside, and a pair of gold sleeve buttons, on which Mr. Allen's name is engraved in full. Mr. A. values them very highly, not for their intrinsic worth, but because they were presents from near and dear friends. Any person seeing the above described articles in the possession of Indians, will confer a lasting favor upon Mr. A. by informing him of the fact, or bringing the thieves to justice.

   It is not known if any of Allen's articles were ever recovered.

6. Paraphrase of "The Dressmaker's Thrush" by W. C. Bennett.
7. Pittsburgh Manuscript, 335-336.
8. This may refer to Endor, a Canaanite village mentioned in the Bible. Its physical location is widely disputed.
9. This neighbor may have been George Tykel, Henry Mattice, or Silas E. Dennis. "Township 19 North, Range 2 West," unattributed map of Thurston County, Washington, Northwest Micro 912.7977, Washington State Library, Olympia.
10. This is a troublesome reference. "Yellow" Prairie probably referred to Yelm Prairie, which was several miles southeast of Olympia proper; a road from Yelm Prairie to the Sound would not have come anywhere near Allen's claim. His claim was situated near the north end of today's Cooper Point peninsula, three miles north of Olympia; in Allen's time, only one or two settlers lived north of him, and their residency would not have warranted the building of a road south to town. Allen did purchase some acreage in Olympia proper, which might have been affected by a road from Yelm Prairie, but the purchase was not made until after this letter was written.
11. Allen termed this the "Hudsons Bay mail canoe" in his Pittsburgh Manuscript, 337.
12. James Fenimore Cooper, who wrote *The Last of the Mohicans*, among other books.
13. Pittsburgh Manuscript, 337-338.
14. Edward Jay Allen to William Allen, July 20, 1853, Yale Scrapbook.

Chapter 6: Building the People's Road

1. The *Columbian* newspaper in 1853 often referred to the Naches Pass wagon road as the "People's Road." This name quickly became common usage, although at times it was called the Cascade Road or the Puget Sound Emigrant Road.
2. Meeker, *Pioneer Reminiscences*, 135.
3. Quoting George Himes, Meeker credited this story to James Longmire and W. O. Bush, whom Himes said was one of the road workers. All of Allen's road builders were working on the road up to October 3 and many of them days longer. Furthermore, one of Himes' sources, W. O. Bush, did not appear in any of the contemporary lists of the road workers. Meeker, *Pioneer Reminiscences*, 136.
4. Theodore Winthrop, author of *The Canoe and the Saddle*, arrived in the Northwest by ship at the end of April 1853. He explored Puget Sound by canoe, traveled over the Cascades via Naches Pass with his Indian guide Loolowcan on his way to Salt Lake, and wrote a draft about his adventures. Winthrop was killed in the opening days of the Civil War and became a northern hero. *The Canoe and the Saddle*, published after Winthrop's death, became an instant best seller and is today considered a Northwest classic. Allen wrote a short piece for the appendix of a 1913 reprint of the book.
5. Edgar and Kirtley.
6. U.S. Army Captain George B. McClellan.
7. The $20,000 Congressional appropriation for building a military road from Puget Sound over the Cascade Mountains was handed over to McClellan, who was charged with the task of road-building.
8. Among the toasts that day was the following by Allen: "THE BACHELORS—If they cannot acquire the chickamines [Chinook jargon for money] with which to return to the States for the girls they left behind, may they be able to supply their demand from among those present." "Proceedings on the Fourth." *Columbian,* July 9, 1853, 2.
9. The Olympia newspaper also reported on the July 9 meeting in which it was determined that due to the non-arrival of Captain McClellan, the citizens of Olympia would have to build the road themselves. Accordingly a committee was created to collect subscriptions to pay for the project. "Road to Walla Walla." *Columbian*, July 16, 1853, 2.
10. Employed as a messenger for an express company, Captain John Goldsbury Parker arrived in Olympia in 1853 carrying the first express mail from San Francisco to Puget Sound. The first session of the territorial legislature was held on the second floor of his mercantile building. Governor Stevens appointed him the territory's first express messenger. In later life he piloted steamboats on Puget Sound. Bennett, 71-72.
11. "Receipts." *Pioneer and Democrat,* September 30, 1854, 3.
12. Weldon Rau, *Pioneering the Washington Territory* (Tacoma: Media Productions Associates, 2003), 7-10.
13. Meeker, *Pioneer Reminiscences*, 139.
14. William Allen was working as a clerk on the Mississippi River steamboat *Norma* based in New Orleans. Edward Jay Allen to William Allen, November 3, 1853, Yale Scrapbook.
15. Allen's reference to the "Great Jehovah" is a paraphrase of the demand Ethan Allen of the Green Mountain Boys made when he captured Fort Ticonderoga from the British during the Revolutionary War. The Olympia newspaper reported that Kirtley's crew left on July 19 to begin work on the eastern part of the road and that Allen's crew was leaving July 23 to begin

work on the western part and predicted that the road would be open in twenty-five days. "The Cascade Road," *Columbian,* July 23, 1853, 2.

16. Allen outfitted in Steilacoom, courtesy of Lafayette Balch, who billed the road crew at cost for their equipment. The eighteen men and their Indian guide (Quiemuth, according to Meeker), riding horses supplied by Leschi (again according to Meeker), rode down to Packwood's Nisqually ferry. The billing stated Allen was charged fifty cents for the use of the ferry. "Cascade Road Fund," *Pioneer and Democrat,* September 24, 1854, 4.
17. Charles Wilkes of the U.S. Navy led the U.S. Exploring Expedition into Puget Sound in 1841. Members of the party followed the Indian trail over Naches Pass. Allen referred to Wilkes' official report, which was being serialized in the Olympia newspaper at the time.
18. Two competing versions of the city of Steilacoom existed in 1853: the port or lower town founded by Lafayette Balch and the upper town founded by John Chapman. Allen said the U.S. Army was paying the HBC $7,000 a year for the use of Fort Steilacoom. Gary Reese stated the annual rent paid to the HBC in 1857 was $600. Perhaps Allen was exaggerating, or perhaps this number is what he was told at the time. Gary Fuller Reese, *A Documentary History of Fort Steilacoom, Washington, Second Edition* (Olympia: Thurston County Historic Commission, 1988), 113.
19. A subsidiary of the Hudson's Bay Company.
20. Pittsburgh Manuscript, 346.
21. From "Siege of Corinth" by Lord Byron.
22. Pittsburgh Manuscript, 346-347.
23. The Minié ball was a type of bullet with a conical point and ridged base, developed by Frenchman Claude Minié. The ridges of the bullet fit into the spiral rifling inside the gun barrel, allowing the ball to travel faster and truer than a bullet with no ridges. "Claude-Étienne Minié," *Encyclopedia Brittanica.*
24. Allen's Chinook Jargon dictionary lists "Ohehut" as "road."
25. Pittsburgh Manuscript, 348-353.
26. Latin for "an event provoking war."
27. Pittsburgh Manuscript, 353-356.
28. Robert S. Moore. "Cutting the Naches Pass," *The Pioneer History of Enumclaw* (Enumclaw, WA: The Woman's Progressive Club, 1941), 61-64.
29. Meeker, *Pioneer Reminiscences,* 151.
30. Theodore Winthrop, *The Canoe and the Saddle* (Tacoma: John H. Williams, 1913), 323-326.
31. Winthrop, 127.
32. Mrs. Lou Palmer, "Narrative of James Longmire, A Pioneer." *Washington Historical Quarterly, Vol. 23, No. 1* (Seattle: Washington University, January 1932). Continued in Vol. 23, No. 2, April 1932), 59.
33. "The Cascade Road," *Columbian,* August 13, 1853, 2.
34. Allen's crew somehow missed the return of the Kirtley party. Allen likely learned that Kirtley had gone back to the settlements from Lieutenant Hodges when he reached Allen's camp on August 29 on his way from the Yakima River area to Steilacoom. Andy Burge, however, told Meeker in his 1904 letter that Allen learned the unwelcome news from a person named Ehformer. Meeker, *Pioneer Reminiscences,* 151.
35. "James Longmire, Pioneer." *Tacoma Sunday Ledger,* August 21, 1892, 9-10.
36. Palmer, "Narrative of James Longmire" Vol. 23, No. 1, 58-60.
37. Palmer, "Narrative of James Longmire," 138.
38. Meeker, *Pioneer Reminiscences,* 136.
39. Moore, 88-91.
40. Winthrop, 288.
41. Winthrop, 93.
42. Robert Cantwell, *Hidden Northwest* (Philadelphia and New York: J. B. Lippincott Co., 1972), 152.
43. Winthrop, 142-144.
44. Lieutenant Hodges probably told him this.
45. Probably a paraphrase of "Good-Bye" by Ralph Waldo Emerson.
46. Unknown source.
47. This was obviously not 4,000 miles as the crow flies.
48. Allen's father was a lead contractor in the construction of the summit railroad tunnel through the Allegheny Mountains of Pennsylvania. William Allen was working on a Mississippi River steamboat. George Allen was lying in his grave in Little Rock, Arkansas.
49. Paraphrase of "Edwin and Angelina" by Oliver Goldsmith.
50. Allen most likely referred to Pyramid Peak, which stands just to the north of Naches Pass. The following year Allen planted a large American flag on its summit, which the Ebey wagon train noted. Today a rough trail leads to the bare summit, which provides a grand view of the route all the way to the prairies of Enumclaw.
51. The *Columbian* noted on August 20, "Moore will go upon the road in a few days with a party of men to give the finishing touch to the whole job." A. W. Moore started out from Olympia just after Kirtley's return and stayed with Allen's men until September, when he went east of Naches Pass to track down Captain McClellan. Moore returned to Olympia immediately after meeting McClellan. He did not go east to help guide in the expected Longmire/Biles

wagon train as Allen suggested he would. Moore, knowing that Nelson Sargent had already gone east to meet the inbound wagon train, likely decided after his meeting with McClellan that there was no need for him to do the same.

52. Umatilla River near Pendleton, Oregon.
53. Some Oregonians supposedly told the inbound emigrants that only part of the road was open and that the plan was to get them started into the mountains and then make them cut their way through, thus creating a road where none existed. "Oregon Alarmed," *Columbian,* August 6, 1853, 2.
54. Pittsburgh Manuscript, 355-356.
55. George Himes, *Transactions of the 35th Annual Reunion of the Oregon Pioneer Association Containing the Annual Address by George H. Himes.* (Portland: Oregon Pioneer Association, June 1907), 146. Susan Biles Drew, "Oregon Trail," Oregon State Historical Society document, 6-7.
56. "Van Ogle's Memory of Pioneer Days." *Washington Historical Quarterly, Vol. 8, No. 4* (Seattle: Washington University, October 1922), 269-270.
57. "Meeker Again Heeds Call of Old Oregon Trail." *Tacoma Ledger*, April 2, 1923, Magazine Section: 4.
58. Meeker, *Pioneer Reminiscences*, 151-152.
59. Oregon Trail emigrants arriving at The Dalles, Oregon, had a choice to either float down the Columbia River to Portland or to follow a land route around Mt. Hood on the Barlow Toll Road. Laurel Hill was the worst section of that route, an extremely steep, rocky, and dangerous chute down which the wagons were lowered with ropes winched around trees to brake the descent.
60. Meeker, *Pioneer Reminiscences*, 145-146.
61. The names are found in "The Cascade Road," *Columbian*, September 24, 1853, 2. A second listing of Allen's work crew may be found in "Tribute to Edward J. Allen, Esq.," *Columbian*, October 8, 1853, 3, which adds the name of Jesse Boyes. The list of Kirtley's men comes from "The Cascade Road," *Columbian*, July 23, 1853, 2.
62. Moore, 90.
63. Chinook jargon for an expression of doubt. Allen probably meant that the third sample was a coal-like substance, but he was not completely certain of its identification.
64. Unknown source.
65. From "The Lytell Geste of Robin Hood," a medieval English ballad.
66. The herring incident occurred on Allen's journey over the Oregon Trail in 1852. Allen was about to eat his last bit of food, a dried herring, when he looked up into the eyes of a starving emigrant. The herring never made it to Allen's mouth. Aiken must have been a sorry sight indeed to recall such hardships. Larsen and Johnson, 216.
67. David Longmire listed three Aiken brothers in the wagon train: James, John, and Glen. The 1853 sources mention no first name. However, in his Pittsburgh Manuscript Allen said that "Jas [James] Aiken was the first one to enter leaving his train back at Wenass and after seeing the provisions sent forward by me pushed on to Olympia." Pittsburgh Manuscript, 357; David Longmire. "First Immigrants to Cross the Cascades," *Washington Historical Quarterly, Vol. 8, No. 1*, (Seattle: Washington University, January 1917), 22-28.
68. Pittsburgh Manuscript, 356-357.
69. Moore, 91.
70. Mat K. Smith took over as the editor of the *Columbian* on October 2, 1853.
71. "The Peoples Road," *Columbian,* October 1, 1853, 2.
72. Meeker, *Pioneer Reminiscences*, 136.
73. Palmer, "Narrative of James Longmire, A Pioneer," 47-48.
74. Five other members of the Longmire/Biles wagon train penned their reminiscences well after the event—David Longmire, Erastus Light, George Himes, Van Ogle, and Susan Isabel Biles Drew. Over the years, many historians have based their accounts primarily on these and James Longmire's reminiscences.
75. Allen started his calendar May 21 (the beginning of the road survey) and ended it on October 8.
76. From "Childe Harold's Pilgrimage" by Lord Byron.
77. On September 10, 1853, the *Columbian* reported that Lieutenant Hodges "has been authorized to offer equivalent to three hundred dollars to anyone who, within two months, will successfully pioneer a train of ten wagons across the mountains." No record has been found of Nelson Sargent, James Longmire, or anyone else ever collecting the money.
78. "Mr. Sargent has reached home again," *Columbian,* October 22, 1853, 2.
79. Ezra Meeker to Edward Jay Allen, January 4, 1905, Meeker Papers, Box 6, Vol. 1.
80. "Great Mass Meeting. The Peoples Road." *Columbian,* October 8, 1853, 2.
81. "Liberality," *Columbian,* October 15, 1853, 2.
82. The *Columbian* on September 24 reported, "Mr. Allen has lost his house and some $2000 worth of timber by fire, while out on this road, hard at work."
83. Mark Tapley, a character in Charles Dickens' novel *Martin Chuzzlewit,* strove to remain jolly at all times and looked for difficult circumstances in life to test his good spirits.

84. From "A Psalm of Life" by Longfellow.
85. Unknown source.
86. Another Dickens character.
87. This census was taken in the late summer and fall of 1853 by United States Marshal J. Patton Anderson. He listed the territorial population as 3,965, including 1,682 males eligible to vote. According to the Washington State Archives, the details of this census have been lost. The 1850 U.S. Census listed the population of "northern Oregon" as 1,049, giving an idea of the size of the immigration over the intervening three years.
88. Paraphrase of "The Lytell Geste of Robin Hood," a medieval English ballad.
89. An account of this adventure may be found in Blankenship, "Biography of William Mitchell," 142-146.
90. Governor Stevens' report to Congress noted a November 3, 1853, snowfall of four inches at Naches Pass.
91. This was the St. Joseph Mission at the Ahtanum, located east of present-day Yakima.
92. Blankenship, "Biography of William Mitchell," 144-145.
93. Dorothy Winston, *Early Spanaway* (Tacoma: Graphic Press, 1976).

Chapter 7: The Captain and the Colonel

1. Winthrop, 328. Allen wrote about McClellan, "I had a certain intimacy with Capt. (by brevet I believe) McClellan, arising from my acting as his secretary for some months, and the intimacy that would come from occupying the same cabin in Olympia." Gibbs is identified as the third roommate on page 327.
2. Gibbs wrote extensively about the Northwest Indians and other topics, drew sketches, and left historians with valuable primary materials. David I. Bushnell, "Drawings by George Gibbs in the Far Northwest, 1849-1851," *Smithsonian Miscellaneous Collections, Vol. 97, No. 8* (Washington, DC: Smithsonian, 1938), Introduction.
3. Allen became a colonel during the Civil War.
4. The Army brevetted officers as a way of honoring them for distinguished service without actually increasing their rank or pay. McClellan's rank was second lieutenant of engineers, brevet captain, U.S. Army.
5. "Col. Allen and His Indian Password," *Washington Herald,* November 5, 1913.
6. Winthrop, 329-330.
7. Pittsburgh Manuscript, 298-299.
8. Philip H. Overmeyer, "George B. McClellan and the Pacific Northwest," *Pacific Northwest Quarterly, Vol. 32* (Seattle: University of Washington, January 1941), 12.
9. Congress may have insisted on calling this a "military" road, but Davis clearly saw a different purpose. Oscar Winther, "Inland Transportation and Communication in Washington 1844-1959," *Pacific Northwest Quarterly, Vol. 30* (Seattle: University of Washington, January-December 1939), 375.
10. The U.S. Army post, originally named Columbia Barracks, was built in 1849 on a rise overlooking the Hudson's Bay Company's Fort Vancouver on the Columbia River. In 1860 the HBC abandoned its fort and the U.S. Army co-opted it, renaming the combined complex Fort Vancouver. Years later it changed names again to Vancouver Barracks. In 1996 the barracks became part of the Fort Vancouver National Historic Preserve.
11. Overmeyer cited his source as U.S. Senate documents *Pacific Railroad Reports*, I, 188. Overmeyer, 20.
12. As mentioned in Chapter 6, Kirtley's men received little recognition from the citizens of Olympia for their efforts in road building east of the Cascade crest. This is perhaps why. Overmeyer, 33.
13. Overmeyer, 47.
14. Ibid, 49.
15. Ibid, 50.
16. Ibid, 52-53.
17. Overmeyer, 58, note 231.
18. "Capt. McClellan." *Pioneer and Democrat,* March 11, 1854.
19. "Shipping." *Weekly Oregonian,* March 25, 1854.
20. Edward Jay Allen to William Allen, January 10, 1854, Yale Scrapbook.
21. David Wilma, "Stevens, Isaac Ingalls (1818-1862)," HistoryLink.org Essay 5314.
22. This meeting took place in mid-December 1853, as Allen was exploring Puget Sound via whaleboat during the first half of December.
23. Corcoran & Riggs was one of the more prosperous banking firms in the United States at the time. A brief history of the firm may be found in "A Washington Philanthropist, Sketch of William W. Corcoran," *Appletons' Journal of Literature, Science, and Art, Volume Eleventh,* (New York: D. Appleton and Company, January 3 to June 27, 1874), 9-11.
24. Edward Jay Allen to William Allen, January 10, 1854, Yale Scrapbook.
25. "About the Cascade Road," *Washington Pioneer,* December 24, 1853, 2.
26. *Journal of the U.S. House of Representatives,* December 13, 1860, 73.

Chapter 8: Whaleboat Voyage

1. "Olympia Bakery," Advertisement. *Columbian*, March 26, 1853, 3; Bancroft, 114, 250; "Weed." Obituary. *Daily Olympian*, January 13, 1896.
2. Don Trosper and Janet Haag, *The History of Tumwater, Vol 2, New Market* (Tumwater, WA: Tumwater Historical Association, 1994), 34-35.
3. Paraphrase of "Benedicite" by John Greenleaf Whittier.
4. Allen probably christened the boat with this name.
5. From "On Ship-Board" by Theodore S. Fay.
6. From "The Sea Witch" by D. W. Rockwell.
7. Richard Hakluyt's *Voyages*.
8. May refer to black walnuts or hickory nuts.
9. Allen's lengthy discourse on historic Northern Pacific explorations has been omitted.
10. Efforts to locate this map have so far been unproductive, if indeed it still exists.
11. Smith's Bay does not appear on present-day maps. Given Allen's description of it being on the right-hand shore, between his claim and Johnson's Point, possible locations include Gull Harbor, Zangle Cove, Little Fish Trap, Big Fish Trap, or most likely Henderson Inlet.
12. The correct name is Johnson Point, which is only about nine miles north of Olympia as the crow flies. The site was named for J. R. Johnson, M.D., who lived in a log cabin on the point, and seems to have dispensed prescriptions mainly of the alcoholic variety. The point was first named in 1841 by Commander Wilkes as Point Moody, for Quartermaster Moody of his command. Hitchman, 139.
13. Mt. Olympus is the tallest peak in the Olympic Mountains, elevation 7,965 feet. British explorer John Meares named the peak in 1778. However, Olympus is not visible from Puget Sound. Allen was looking at either Mt. Anderson or Mt. Constance, which are the two highest Olympic peaks visible from the Sound. The Olympic Mountains were not thoroughly explored until the 1890s. "European and Euro-American History," Olympic National Park.
14. Mt. St. Helens.
15. Mt. Rainier, tallest peak in the Northwest, today is measured at 14,410 feet. "Mount Rainier," USGS Volcano Hazards Program.
16. At changes of tide, all the water coming to or from southern Puget Sound must pass through the Narrows, a mile-wide, four-mile-long channel known for its strong rip tides. The name was first given to the area by the Wilkes Expedition of 1841. Hitchman, 301.
17. Lieutenant Ringgold of the 1841 Wilkes Expedition was tasked with exploring the inland waters of the Sound, and the bay where he commenced his mission was called "Commencement Bay." Today, the bay is surrounded by the city of Tacoma. Hitchman, Addendum, 2.
18. The Puyallup River begins as meltwater from the Puyallup and Tahoma glaciers on Mt. Rainier. Hitchman, 243.
19. John M. Swan was born in Scotland circa 1823, and became a naturalized U.S. citizen in Thurston County in 1855. He built the first schoolhouse in Olympia; settled a land claim which he later platted and named Swantown; and on censuses was listed variously as a nurseryman, pomologist (specialist in apples and pears), and shipbuilder. "Swan," Obituary. *Daily Olympian*, January 13, 1904, 3.
20. Charles W. Riley was born in New York circa 1823, and settled on a land claim in Pierce County in 1853. He served as a captain in the East Puyallup Rangers during the Indian uprising of 1855-56, and was described on a muster roll as five feet eleven inches tall, with light hair, gray eyes, and a light complexion. "Military Department…Charles W. Riley." He was listed as a merchant on the 1860 census.
21. The Duwamish River in 1853 began at the confluence of the Green and Black Rivers. The combined flow wound north to Elliott Bay, the present-day name for Allen's Duwamish Bay. Elliott Bay is now surrounded by the city of Seattle. The view is not quite as extensive as Allen stated. Three Tree Point (also known as Point Pully) is probably the farthest extent of the view from Commencement Bay north.
22. Mt. Baker, 10,778 feet, the northernmost of the volcanoes in Washington's Cascade Range, lies about thirty miles east of Bellingham. The name honors Lieut. Joseph Baker, a member of Captain George Vancouver's exploration party of 1792. The local Nooksack Indians called the mountain "Koma-Kulshan," meaning "white, shining mountain." Hitchman, 194.
23. This fourteen-mile long island was named by Captain George Vancouver for Captain James Vashon of the Royal Navy. Hitchman, 318.
24. Admiralty Inlet is "the portion of northern Puget Sound which extends northward from Foulweather Bluff to the Strait of Juan de Fuca, between Whidbey Island on the east, and Kitsap and Jefferson counties on the west." Hitchman, 2.
25. Originally named New York, the settlement of Alki preceded the town of Seattle, was founded in 1851 by the Denny party, and was platted by C. C. Terry after most of the settlers had moved to the east side of Elliott Bay. The Chinook word Alki means "by and by" or "soon."

"Alki" became part of the Washington State motto. Hitchman, 4. (Alki is correctly and historically pronounced "al-kee," although most Washington residents now pronounce it with a long "i.")

26. Seattle, now Washington's largest city, lies on the shores of Elliott Bay. It was named for Chief Sealth, who was very friendly to white settlers. Hitchman, 267.
27. Charles C. Terry was born in New York in 1830, went to California in 1849, and came to the Northwest aboard the schooner *Exact* in 1851. He ran a sawmill and opened the first mercantile in King County. He first established the village of Alki, but later moved to Seattle. He married Mary Jane Russell in 1856, and donated land for the University of Washington. He died of tuberculosis in 1867. "Charles C. Terry," University of Washington Libraries Digital Collections.
28. Paraphrase of "Hampton Beach" by John Greenleaf Whittier.
29. Their campsite was on today's Alki Beach in West Seattle.
30. Allen eventually purchased three acres in Seattle, from Carson Boren, near the present intersection of Seneca Street and Terry Street. *King County Deeds, Volume AC,* 60, and *Volume ABC,* 47, King County Archives, Seattle. Edward Jay Allen to Elizabeth Allen, April 17, 1861, Cahill Collection.
31. Wilkes named these high clay cliffs Skadg-it Head, but the official name is now Scatchet Head. Hitchman, 266.
32. From "The Corsair" by Lord Byron.
33. Allen no doubt meant Swan and Riley's establishment at Commencement Bay.
34. Allen probably meant Snohomish River, which does indeed empty into Puget Sound just a few miles below (north of) Scatchet Head. The name is a corruption of the name for the local Indian tribe, which has been linguistically rendered as "Sda-hob-bish," meaning "tidewater people." Hitchman, 279.
35. Paraphrase of "By That Lake, Whose Gloomy Shore" by Thomas Moore.
36. Paraphrase of "An Ode" by Charles Sprague.
37. Named in honor of Lord Hood, a member of the British Board of Admiralty, Hood Canal is about eighty miles long. Its fishhook shape runs between Kitsap and Mason counties, and then bends sharply to the east. Hitchman, 126.
38. David S. Maynard was born in Vermont in 1808, and came to the Pacific coast in 1850. He was a medical doctor, and first settled in Olympia, then traveled to California. He returned to the Northwest, and moved his base of operations to the fledgling town of Seattle, where he opened a hospital. Although he had several personal and professional misfortunes, he is generally credited with being one of the founders of the metropolis of Seattle. He died there in 1873. Junius Rochester, "Maynard, Dr. David Swinson," HistoryLink.org.
39. Snoqualmie Falls, a very impressive and picturesque waterfall which drops 268 feet. The Snoqualmie and Skykomish rivers meet and form the Snohomish River. Hitchman, 279.
40. Penn Cove is a harbor on the east shore of Whidbey Island, named in 1792 by Captain George Vancouver for a "particular friend." In 1848, Thomas Glasgow took up a donation land claim outside present-day Coupeville. Glasgow, however, became frightened of the Indians and left for points south. Thomas Ebey came in 1850, and from that time on, more and more settlers arrived, taking up permanent residence. Hitchman, 227-228.
41. A small town, established circa 1850, on the west end of Penn Cove. Hitchman, 58.
42. The first European to settle on Whidbey Island was Father Francis Blanchet. He arrived in 1840 and stayed with the native residents for about 11 months, performing many baptisms and establishing a mission. Cook, 13.
43. Paraphrase of "A Vision of the Hudson" by William Cox.
44. This name led the authors on a merry chase. Early censuses, donation land claim records, pioneer accounts, and myriad old maps contained nothing remotely similar to Schroreckbin. Finally, a clue turned up in "Indian Nomenclature and Localities in Washington and Oregon" written by George Gibbs in 1853. Gibbs, who compiled native names for many locations in the Northwest, noted the following: "Cho-mahlst or S'cho-bahlst, S.E. Pt Penn's Cove." If Allen's letter used the name S'cho-bahlst Point, Pittsburgh typesetters may have mangled it into Schroreckbin's Point. Unfortunately, Allen's Pittsburgh Manuscript did not contain a description of the whaleboat voyage, so the authors were unable to use it to corroborate their guess. Upon personally examining the geography of the area, the authors have settled on present-day Snakelum Point as the most likely suspect for the point Allen describes. Below (south of) the point is the extensive Smith Prairie, and just to the east is a high bluff—both mentioned by Allen. The point contained an Indian village that was abandoned in the early 1850s following smallpox epidemics.
45. Probably Joseph S. Smith, who located on Smith's Prairie in 1853. Smith had come to Oregon Territory

some years earlier, lived in the Portland area, and served as an early Portland mayor. He also served in the Washington Territorial Legislature, and was District Attorney for the Territory for two years. He later moved back to Portland, where he was a representative to Congress in 1870-71. Cook, 55.

46. From "Lalla Rookh" by Thomas Moore.
47. Unknown source.
48. Paraphrase of "An Autumn Thought" by Bayard Taylor.
49. William B. Wilton was born circa 1820 in Boston, arrived in Oregon Territory in December 1850, and settled his claim near Steilacoom shortly thereafter. *Washington Territory Donation Land Claims,* 98.
50. See Chapter 1, page 6.
51. Weed did indeed purchase this tract; his name appears on survey maps of the area.
52. The *Fairy* was a small sidewheeler brought to Puget Sound from San Francisco on the deck of a sailing vessel in 1853. She was the first steamer on the Sound to engage in purely local business. Captain Warren Gove was her first master. In October 1857, while docked in Steilacoom, her boiler exploded, and the hull sank. Wright, 45.
53. In 1853, the following advertisement appeared in several issues of Olympia's *The Pioneer*: "WANTED!—Twenty coal miners, at Marmosa, Bellingham Bay. Wages from Sixty to Seventy dollars per month, or one dollar per ton and found. W. A. Howard, Superintendent, FSCM Association."
54. Either Hartstene Island or Squaxin Island. Hartstene is by far the larger of the two, but Squaxin lies more directly across the mouths of the inlets Allen mentioned. Squaxin Island today comprises the Squaxin Indian Tribe reservation.
55. This may be a misprint of "*cum multis aliis,*" which could mean "among many others."
56. Unknown source.
57. Edward Jay Allen to William Allen, December 27, 1853, Yale Scrapbook.

Chapter 9: Allen for the Senate

1. The Whig Party, formed originally around opposition to Andrew Jackson's presidency (1829-1837), operated as the second political party in the U.S. until the 1840s when it began fracturing over the question of slavery. The party favored protective tariffs, internal improvements, the supremacy of the Congress over the Executive Branch, and a growth-oriented monetary policy. Some well-known Whigs were Daniel Webster, William Henry Harrison, Zachary Taylor, and Abraham Lincoln. "Whig Party," *Encyclopedia Britannica*. The Olympia Whig convention was held December 17, 1853. "To-Day!" *Washington Pioneer*, December 17, 1853, 2.
2. The Monticello Convention. See Chapter 2.
3. See Chapter 3, note 8. The Olympia Free Soil Party held a convention in Olympia on August 19, 1854. Allen did not attend, as he was in the Cascade Mountains working on his road.
4. From "Be Firm!—Be True!" by W. D. Gallagher, 1847.
5. The Allen family was ardently anti-slavery. Edward, upon his return to Pittsburgh, became involved in the Kansas Immigrant Aid Society prior to the bloody miniature civil war in that future state. "I am one of the board of managers for the Kansas Aid Association with Eddie which meeting every day keeps me busy." William Allen to Rebecca Allen Turner, March 17, 1856, Turner Collection, UALR; and "Kansas Affairs," *Cleveland Herald*, June 20, 1856, 3.
6. Unknown source.
7. Edward Jay Allen to William Allen, January 10, 1854, Yale Scrapbook.
8. "Whig Nominations." *Washington Pioneer,* December 31, 1853, 2.
9. Winthrop, 328.
10. "Proclamation," *Pioneer and Democrat,* March 4, 1854, 2.

Chapter 10: Back to the Cascades

1. Rebecca Allen to William Allen, January 31, 1854, Rosanio Collection.
2. Rebecca Allen to Edward Jay Allen, February 6, 1854, Rosanio Collection.
3. Edward Jay Allen to William Allen, January 10, 1854, Yale Scrapbook. Allen corresponded with Stevens after his return to Pittsburgh. In the Isaac Stevens papers on microfilm at the Washington State Library is a chatty letter from Allen to Stevens dated 1859. About the same time, Stevens addressed a letter to Allen's young niece, Mary Fundenburg; only the envelope survives. In 1861, Stevens endorsed a letter of recommendation written by General McClellan requesting an officer's commission for Allen.
4. Winthrop, 328.
5. *Pittsburg Dispatch*, May 20, 185[4], Yale Scrapbook.
6. Winthrop, 330.
7. Meeker, *Pioneer Reminiscences*, 114.
8. Isaac I. Stevens, *Narrative and Final Report of Explorations* (Washington, DC, 1860), 190.
9. "Lieut. Arnold," *Pioneer and Democrat*, May 20, 1854, 3.

10. George Gibbs, "Journal No II 1854-1855, Cascade Road—Indian Notes," Northwest Boundary Records of the U.S. Commissioner (National Archives, College Park, MD).
11. "Mr. Allen mistook the directions from here as to this and cut his road to the left, thereby much lengthening it." Gibbs, "Journal," July 4.
12. Allen Porter's residence near present-day Enumclaw.
13. Father Pandosy established a Catholic mission in the Kittitas valley in 1848; ill health forced him to abandon the mission the following year. On April 3, 1852, he and Father D'herbomez founded Saint Joseph's Mission at Ahtanum Creek in the Yakima Valley. In 1854 Father D'herbomez was reassigned to Olympia. Paula Becker, "Saint Joseph's Mission," HistoryLink.org. Perhaps Father Pandosy was on a trip over the mountains relating to this reassignment.
14. Also called kinnikinnick or bearberry, *Arctostaphylos uva-ursi* is a creeping evergreen shrub that was commonly used by fur traders and perhaps Native Americans to extend their tobacco supply. Hitchcock and Cronquist, 342; Lewis J. Clark, *Wild Flowers of British Columbia* (Sidney, BC: Gray's Publishing Limited, 1973), 360. Today it is a prized ornamental.
15. See Chapter 4, note 60, for an account of LeTete. A good view of the peak can be had from Highway 410 heading west into the village of Greenwater. The summit has been logged giving it much the same appearance as it had in 1853-54.
16. This was most likely Twenty-eight Mile Creek.
17. Gibbs, "Journal," July 11.
18. Gibbs, "Journal," July 15.
19. According to Gibbs' journal James Hurd, the financial manager of the 1853 road fund, was at the Naches summit on July 15, five days after he left Olympia, demonstrating the utility of Allen's roadwork. Ebey's diary on September 3 said, "This forenoon we met on the road, Mr. Jas. K Hurd of Olympia W.T. He has been on the road all summer giving the emigrants information about the new road from Walla Walla to the Sound—Hearing of our train he came out to meet us." Susan Badger Doyle and Fred Dykes, eds., *The 1854 Oregon Trail Diary of Winfield Scott Ebey* (Independence, MO: Oregon-California Trails Association, 1985), 174. While Hurd was out on the road he heard through the Oregon Trail grapevine that Ezra Meeker's kinfolk were delayed and short of provisions. He sent a message to Meeker advising that he should go to their assistance and guide them over Naches Pass. Meeker filled chapters of his book *Pioneer Reminiscences* in relating that adventure.
20. "Road across the Cascades," *Pioneer and Democrat*, July 8, 1854, 2.
21. "Messrs. Ensign, Kirtley and Blankenship," *Pioneer and Democrat*, August 19, 1854, 3.
22. Himes, "Very Early Ascents," 199.
23. "Aged Settler," *The Portland Evening Telegram*, November 11, 1909, 4.
24. Fred Becky, *Cascade Alpine Guide Climbing and High Routes 1: Columbia River to Stevens Pass, Second Edition* (Seattle: The Mountaineers, 1987), 56.
25. This is the famous cliff discussed in Chapter 6.
26. Winthrop, 328-329.
27. Meeker, *Pioneer Reminiscences*, 97-98.
28. Meeker, *Pioneer Reminiscences*, 116.
29. "Mr. Silas Galliher," *Pioneer and Democrat*, September 2, 1854, 2.
30. "Arrival of Immigrants," *Pioneer and Democrat*, September 16, 1854, 1.
31. Doyle and Dykes, 195.
32. Goodell, 126.
33. Doyle and Dykes, 196.
34. Doyle and Dykes suggest Kelly Butte as Ebey's Mount Ike. Pyramid Peak is a more likely choice as it adjoins Naches Pass via a short two-mile ridge, has a rocky summit on which a flag would stand out, and is the most prominent peak visible from the ridge the wagon train descended after leaving the pass. Doyle and Dykes, 197.
35. Ibid.
36. Ibid, 197-198.
37. Ibid, 201-202.
38. The steamer *Major Tompkins* was built in Philadelphia in 1847. She served in New Orleans and New York before she headed for the Pacific where she plied the Sacramento River. In 1854 Captain James M. Hunt and John H. Scranton, a San Francisco merchant, brought the *Tompkins* to Puget Sound to serve a mail contract. Wright, 52. She was greeted with gunfire and shouts of joy at each stop. The city of Steilacoom, being frugal with gunpowder, greeted the *Tompkins* by blowing up the town's stumps. "List of Passengers." *Pioneer and Democrat*, December 23, 1854, 3.
39. Blanche Billings Mahlberg, "Edward J. Allen, Pioneer and Roadbuilder," *Pacific Northwest Quarterly, Vol. 44, No. 4* (Seattle: University of Washington, October 1953), 160.
40. William Billings was a neighbor and friend of Allen. He lived just across Budd Inlet from Allen's cabin. Blanche Billings Mahlberg was William Billings' daughter.

Chapter 11: Homeward Bound

1. An obvious printing error, this should read 1855.
2. Another error; the *Halcyon* left Olympia on January 20. "Puget Sound Shipping Report," *Pioneer and Democrat,* January 27, 1855, 3.
3. David Burntrager was born in Ohio, and arrived in the Northwest in 1851. On November 16, he married Matilda in Albany, Oregon Territory. The couple settled their claim, just north of Allen's land, on October 18, 1853. *Washington Territory Donation Land Claims,* 58. According to his obituary, the Burntragers stayed on that claim for some time, then went to Oregon, and finally returned to Thurston County and settled near Bush Prairie (south of present-day Tumwater). In 1893, Burntrager was killed by a falling tree at his home on the Deschutes River. Matilda and one son survived him. "Another Pioneer Gone," *Washington Standard,* July 7, 1893.
4. Allen referred to Johann Goethe's 1788 play *Egmont* in which the heroine Klarchen puts an end to her life when her lover is condemned to death.
5. The schooner *Halcyon* was purchased by Captain George Flavel, who made "money fast in the coasting trade." Wright, 54. His house in Astoria exists today as the Flavel House Museum. During Allen's trip of 1855, however, the *Halcyon* was captained by Henry McDonough, who brought the brig to Olympia in thirteen days from San Francisco. "Late and Interesting News," *Pioneer and Democrat,* January 6, 1855, 2. Captain McDonough was the man who suggested that Allen should "eat a little breakfast" to ease his seasickness.
6. A sailor's stew of meat, vegetables, and hardtack.
7. This steamer was built in 1853 in New York, and was originally named *San Francisco.* She was slated for the Australia-Panama run, and arrived in Panama in 1854. She was soon purchased by the Pacific Mail Steamship Company, and served on the San Francisco-Panama route until 1869. "SS *Golden Age.*"
8. The theatre was built in 1853 and was of a "mixed Renaissance" style. Gas lights were installed in 1854. The building burned in 1857, and was rebuilt by 1861. James R. Smith, *San Francisco's Lost Landmarks* (Sanger, CA: Word Dancer Press, 2005), 99-102.
9. "The Elixir of Love" was a popular opera, written by Donizetti.
10. Soprano Anna Bishop was born in London in 1810. She left her husband and toured Europe, America, and Australia with French harpist and composer Nicolas-Charles Bochsa. Bishop died in 1884. Edward T. James, *Notable American Women, Vol. 1* (Cambridge, MA: Radcliffe College, 1971), 148-149.
11. Born in 1789, Bochsa created a scandal when he absconded to Europe with Anna Bishop. E. J. Lea-Scarlett, *Australian Dictionary of Biography.*
12. Allen's lengthy description of the opera is omitted.
13. *Wearyfoot Common, or The Battle of Life,* by Scottish novelist Leitch Ritchie, was published in 1855. The story may have been serialized in *Chambers's Journal* (a popular monthly magazine published in Edinborough, and edited for some time by Ritchie himself) before it was ever published as a free-standing book. If so, Allen was probably reading an issue of the magazine containing the "few remaining chapters."
14. The banking firm of Page, Bacon & Co. was established in St. Louis, Missouri, by Daniel Page and Henry Bacon. By the beginning of 1855, the firm was on the verge of failure, and on February 15 saw a run on the firm's San Francisco branch. Ralph Moody, *Wells Fargo,* (University of Nebraska: Bison Books, 2005) 68-69. Allen narrowly missed the bank's collapse.
15. The 240-foot steamship *Uncle Sam* was built in New York in 1852, launched in 1853, and by 1855 was owned by the Nicaragua Steamship Company. She served the Nicaragua-San Francisco route, and thereafter operated under a succession of owners. "SS *Uncle Sam.*"
16. Rather than endure a very lengthy sailing trip around Cape Horn, travelers could take a shorter route—sail down either coast to Nicaragua or the Isthmus of Panama, then travel overland to the opposite shore and another boat trip. In very early days, the overland crossing would be made on foot or by mule. Later, a railroad was established.
17. Edward Jay Allen to William Allen, March 1855, Yale Scrapbook.
18. The SS *Prometheus* was a wooden sidewheel steamer, launched in New York in 1850. Stephen Dando-Collins, *Tycoon's War* (Philadelphia: Da Capo Press, 2008), 14.
19. Could this be the same Benjamin Close who served as reverend of the Methodist Church in Olympia during Allen's time there?
20. "Puget Sound Shipping Report." *Pioneer and Democrat,* January 27, 1855.
21. "Postscript. San Francisco Correspondence," Sacramento *Daily Union,* February 17, 1855.
22. "Passengers," *Daily Alta California,* February 17, 1855.
23. "Shipping Intelligence," *Daily Alta California,* February 27, 1855.

Epilogue

1. Allen's literary talent lived on in at least one of his descendants. His grandson, William Hervey Allen, Jr., shortened his name to Hervey Allen, and became one of the best-selling authors of the Depression with his monumental work, *Anthony Adverse*. Hervey was a favored grandson, learned to love literature in his grandfather's expansive home library, and inherited many of his grandfather's belongings.
2. "Edward Jay Allen," *Chronicle Telegraph* (Pittsburgh), December 27, 1915.
3. "Edward Jay Allen," Pioneers' Society of Washington papers, Fiske Genealogical Library, Seattle, WA.

Bibliography

Archival Sources

Hervey Allen Papers. University of Miami, Coral Gables Library.

King County Deeds, Volumes AC and ABC, King County Archives, Seattle.

"Meeker Papers." Washington State Historical Society Research Center, Tacoma, Washington. Features correspondence between Allen and Ezra Meeker.

Military Department, Muster Rolls, 1855–1856, Online 2008, Washington State Archives, Office of the Secretary of State. http://www.digitalarchives.wa.gov.

Pioneers' Society of Washington papers. "Edward Jay Allen." Fiske Genealogical Library, Seattle, WA.

Pittsburgh Manuscript: *Allen, Hervey, Carbon Manuscript of the Oregon Trail by Col. Edward Jay Allen.* University of Pittsburgh Library Special Collections Division. Among the papers of Allen's grandson Hervey Allen was the Pittsburgh Manuscript. Also found in the collection were many of Allen's notes used in preparation of the manuscript, newspaper clippings about his life, obituaries, and other family information, including Allen's own Oregon Trail diary of 1852.

Rosanio Collection. Private collection of Gustave Rosanio, New Jersey. Includes many letters of Rebecca Allen Turner, and another sister, Amelia.

The Turner family papers, 1778–1929 [manuscript]. Rare Book, Manuscript, and Special Collections Library, Duke University, Durham, NC. Holds letters written in the 1840s and 1850s by Allen, his brother William, Allen's father, and his sister Rebecca Allen Turner.

Thurston County Deeds, Volume 1. Washington State Archives, Olympia.

Thurston County First Record Book, 1852–1857, Washington State Archives, Digital Archives, http://www.digitalarchives.wa.gov.

"Turner Family Papers" (UALR.0031). Ottenheimer Library, University of Arkansas Little Rock. Includes many letters and family photographs of Allen's sister, Rebecca Allen Turner.

Washington Territorial Volunteer Papers, Indian War Correspondence, 1855–1857. Compiled by Office of the Secretary of State, Division of Archives and Records Management. Tumwater, WA: Imaging and Preservation Services, Division of Archives and Records Management, 1990.

Yale Scrapbook: "Oregon Trail / Eddie's Letters" Scrapbook. Beinecke Rare Book and Manuscript Library, Yale University. Allen's sister Rebecca's scrapbook containing the 1852–1855 *Pittsburg Dispatch* letters. Scanned images of the scrapbook can be found online at: http://brbl-dl.library.yale.edu/Record/3441271.

We also owe many thanks to Allen descendants Winthrop Baylies and Elizabeth Allen Cahill who supplied us with the wonderful daguerreotype of Allen, along with much other family history.

Manuscripts, Books and Articles

"A Washington Philanthropist, Sketch of William W. Corcoran," *Appletons' Journal of Literature, Science, and Art, Volume Eleventh.* New York: D. Appleton and Company, January 3 to June 27, 1874.

Bagley, Clarence B. *History of Seattle from the earliest settlement to the present time.* Chicago: The S. J. Clarke Publishing Company, 1916.

Bancroft, Hubert Howe. *Washington, Idaho, and Montana: 1845-1889.* San Francisco: History Co., 1890.

Barnard, J. G. *Standard History of Pittsburgh, Pennsylvania.* Chicago: H. R. Cornell & Company, 1898.

Becky, Fred. *Cascade Alpine Guide Climbing and High Routes 1: Columbia River to Stevens Pass, Second Edition.* Seattle: The Mountaineers, 1987.

Bell, Brian H., and Gregory Kennedy. *Birds of Washington State.* Auburn, WA: Lone Pine Publishing International Inc., 2006.

Bennett, Robert A., ed. "Captain J. G. Parker." *A Small World of Our Own. Authentic Pioneer Stories of the Pacific Northwest from the Old Settlers Contest of 1892.* Story written by Captain J. G. Parker, Olympia, Washington, February 10, 1893. Walla Walla, WA: Pioneer Press Books, 1985.

Bergh, Edward. *Betsy and John Edgar: Pioneer Settlers on the Yelm Prairie.* Unpublished, 2009.

Bird, Annie Laurie. *Old Fort Boise.* Parma, ID: Old Fort Boise Historical Society, 1971.

Blankenship, Mrs. George E., ed. "Biography of William Mitchell." *Early History of Thurston County, Washington, Together With Biographies and Reminiscences of Those Identified with Pioneer Days.* Olympia, WA: Publisher unknown, 1914.

Bowden, Angie Burt. *Early Schools of Washington.* Seattle: Lowman and Hanford Company, 1935.

Boyd, Robert. *The Coming of the Spirit of Pestilence.* Seattle: University of Washington Press, 1999.

Branson, C. J. *History of the State of Idaho.* New York: Charles Scribner and Sons, 1918.

Brown, Randy. *Historic Inscriptions on Western Emigrant Trails.* Independence, MO: Oregon-California Trails Association, 2004.

Brown, Randy. "OCTA'S MARKING CONTINUES ALONG THE TRAIL." *News From the Plains: Newsletter of the Oregon-California Trails Association.* Independence, MO: Oregon-California Trails Association, April 1994.

Brown, Randy, and Reg Duffin. *Graves and Sites on the Oregon and California Trails.* Independence, MO: Oregon-California Trails Association, 1998.

Bushnell, David I. "Drawings by George Gibbs in the Far Northwest, 1849–1851." *Smithsonian Miscellaneous Collections, Vol. 97, No. 8.* Washington, DC: Smithsonian, 1938.

Cantwell, Robert. *Hidden Northwest.* Philadelphia and New York: J. B. Lippincott Co., 1972.

Clark, Lewis J. *Wild Flowers of British Columbia.* Sidney, BC: Gray's Publishing Limited, 1973.

Clark, Louis Gaylord, ed. *The Knickerbocker or New York Monthly Magazine, Volume 43.* New York: Samuel Hueston, June 1854.

Cook, Jimmie Jean. *"A particular friend, PENN'S COVE," A History Of The Settlers, Claims, and Buildings Of Central Whidbey Island.* Coupeville, WA: Island County Historical Society, 1973.

"Crossing the Plains in 1852, Mrs. Cecilia Emily McMillan Adams." *Transactions.* Portland: Oregon Pioneer Association, 1904.

Dale, Edward Everett, ed. "Journal of James Aiken, Jr." *University of Oklahoma Bulletin*. Norman, OK: University of Oklahoma, June 1, 1919.

Dando-Collins, Stephen. *Tycoon's War*. Philadelphia: Da Capo Press, 2008.

Daughters of the American Revolution. *Genealogical and historical gleanings, Series 1; Genealogical Records Committee report, Series 2; Family records of Washington pioneers, Series 3*. DAR, Washington State Society, 1927–2000.

Dickey, George, ed. *The Journal of Fort Nisqually Commencing May 30, 1833; Ending September 27, 1859. Section 9*. Tacoma: Fort Nisqually Historic Site, Metropolitan Park District of Tacoma, 1989.

Dietrich, William. *Northwest Passage: the Great Columbia River*. New York: Simon & Schuster, 1995.

Doyle, Susan Badger, and Fred Dykes, ed. *The 1854 Oregon Trail Diary of Winfield Scott Ebey*. Independence, MO: Oregon-California Trails Association, 1997.

Eno, Clara B. *History of Crawford County, Arkansas*. Van Buren, AR: The Press-Argus, undated.

Flagg, Charles Alcott. *Colonial and Revolutionary Families of Pennsylvania: Genealogical and Personal Memoirs*. New York: Lewis Publishing Company, 1911.

Floyd, Ted. *Smithsonian Field Guide to the Birds of North America*. New York: Harper Collins, 2008.

Franzwa, Gregory. *Maps of the Oregon Trail*. St. Louis: Patrice Press, 1990.

Gibbs, George. *Indian Nomenclature and Localities in Washington and Oregon*. 1853-1858. Unpublished notes, Washington State Library.

Gibbs, George. "Journal No II 1854–1855, Cascade Road—Indian Notes," located in Records relating to International Boundaries (Record Group 76). Northwest Boundary Records of the U.S. Commissioner, Journal of Exploring Surveys, 1854-55 (Pl 170, Entry 198.) National Archives, College Park, MD.

Harlow, W. M., and E. S. Harrar. *Textbook of Dendrology, 4th Edition*. New York: McGraw-Hill, 1958.

Himes, George. *Transactions of the 35th Annual Reunion of the Oregon Pioneer Association Containing the Annual Address by George H. Himes*. Portland: Oregon Pioneer Association, June 1907.

Himes, George. "Pioneer Reminiscences." *Transactions of the 53rd Annual Reunion of the Oregon Pioneer Association*. Portland: Oregon Pioneer Association, July 1925.

Himes, George. "Very Early Ascents." *Steel Points, Vol. 1, No. 4*. Portland: William Gladstone Steel, July 1907.

Himes, George. "The Oxen at Naches Pass." *Washington Historical Quarterly, Vol. 14, No. 1*. Seattle: Washington University, January 1923.

History of Seattle, Washington. New York: American Publishing and Engraving Company, 1891.

Hitchcock, C. Leo, and Arthur Cronquist. *Flora of the Pacific Northwest*. Seattle: University of Washington Press, 1973.

Hitchman, Robert. *Place Names of Washington*. Tacoma: Washington State Historical Society, 1985.

Holmes, Kenneth. *Covered Wagon Women, Diaries & Letters From the Western Trails, 1854–1860, Vol. 7*. Lincoln: University of Nebraska Press, 1998.

"Honorary Alumni Directory." *Alumni Directory, University of Pittsburgh, Vol. 2, 1787–1916*. Pittsburgh: General Alumni Association, University of Pittsburgh, 1916.

Hooper, Dean, and Roberta B. Longmire. *Yelm Pioneers and Followers, 1850–1950*. Yelm, WA: Yelm Prairie Historical Society, 1999.

Hoselton, Anna. "The Naches Pass Road." MS, 1987. Washington State Library.

James, Edward T. *Notable American Women: A Biographical Dictionary, Volume I*. Cambridge, MA: Radcliffe College, 1971.

Johnson, Karen L. "Forts Along the Cowlitz." *Cowlitz Historical Quarterly, Volume 52, Number 2*. Kelso, WA: Cowlitz County Historical Society, June 2010.

Johnson, Karen L. "James Gardiner and the Hardbread Hotel." *Cowlitz Historical Quarterly, Volume 52, Number 2*. Kelso, WA: Cowlitz County Historical Society, June 2010.

Johnson, Karen L. "Through the Bowels of the Land." *Cowlitz Historical Quarterly, Volume 51, Number 1*. Kelso, WA: Cowlitz County Historical Society, March 2009.

Johnson, Serene A., and Karen L. Johnson. "Louisa Jackson's Diary of 1865." *The Lewis County Historian, Volume 26, Number 2*. Chehalis, WA: Lewis County Historical Museum, May 2004.

Larsen, Dennis M., and Karen L. Johnson. *Our Faces Are Westward: The 1852 Oregon Trail Journey of Edward Jay Allen*. Independence, MO: Oregon-California Trails Association, 2012.

Larsen, Dennis, and Ken Keigley. *The Meeker Family Over The Oregon Trail and Naches Pass 1854*. Unpublished, 2008.

Leighton, Caroline C. *Life at Puget Sound with Sketches of Travel in Washington Territory, British Columbia, Oregon and California, 1865-1881*. Boston: Lee and Shepard, Publishers, 1883.

Lloyd, James T. *Lloyd's Steamboat Directory*. Cincinnati: James T. Lloyd and Company, 1856.

Longmire, David. "First Immigrants to Cross the Cascades." *Washington Historical Quarterly, Vol. 8, No. 1*. Seattle: Washington University, January 1917.

Lyons, Chester Peter. *Trees, Shrubs and Flowers to Know in Washington*. Toronto: J. M. Dent and Sons Ltd., 1956.

Magnusson, Elsa Cooper. "Naches Pass." *Washington Historical Quarterly, Vol. 25, No. 3*. Seattle: Washington University, July 1934.

Mahlberg, Blanche Billings. "Edward J. Allen, Pioneer and Roadbuilder." *Pacific Northwest Quarterly, Vol. 44, No. 4*. Seattle: University of Washington, October 1953.

McArthur, Lewis A. *Oregon Geographic Names*. Portland: Oregon Historical Society Press, 2003.

Meany, Edmond S. *Living Pioneers of Washington; columns appearing in the Seattle Post-Intelligencer, October 1915–May 1920*. Seattle: Seattle Genealogical Society, c1995.

Meany, Edmond S. *Origin of Washington Place Names*. Seattle: University of Washington Press, 1923.

Meeker, Ezra. *Pioneer Reminiscences of Puget Sound*. Seattle: Lowman and Hanford, 1905.

Meeker, Ezra. *The Busy Life of Eighty-Five Years of Ezra Meeker*. Indianapolis: Wm. B. Buford, 1916.

Miles, Charles, and O. B. Sperlin, eds. *Building a State: Washington 1889-1939*. Tacoma: Washington State Historical Society, 1940.

Moody, Ralph. *Wells Fargo*. University of Nebraska: Bison Books, 2005.

Moore, Robert S. "Cutting the Naches Pass." *The Pioneer History of Enumclaw*. Enumclaw, WA: The Women's Progressive Club, 1941.

Mullan, Capt. John A. "From Walla Walla to San Francisco from the Washington Statesman (Walla Walla) of November 29 and December 6, 1862." *Quarterly of the Oregon Historical Society, Vol. 4*. Portland: Oregon Historical Society, 1903.

Munnick, Harriet Duncan. *Catholic Church Records of the Pacific Northwest, Vancouver and Stellamaris Mission*. St. Paul, OR: French Prairie Press, 1972.

"Notes and Documents; Letters of Governor Isaac I. Stevens, 1853-1854." *Pacific Northwest Quarterly, Vol. 30, No. 3.* Seattle: University of Washington, July 1939.

Officers of the Army and Navy (Volunteer) Who Served in the Civil War. Philadelphia: L. R. Hamersly & Company, 1893.

Overmeyer, Philip H. "George B. McClellan and the Pacific Northwest." *Pacific Northwest Quarterly, Vol. 32.* Seattle: University of Washington, January 1941.

Palmer, Gayle, and Shanna Stevenson, eds. *Thurston County Place Names: A Heritage Guide.* Olympia, WA: Thurston County Historic Commission, 1992.

Palmer, Mrs. Lou. "Narrative of James Longmire, A Pioneer." *Washington Historical Quarterly, Vol. 23, No. 1.* Seattle: Washington University, January 1932. Continued in Vol. 23, No 2, April 1932.

Palmquist, Peter E., and Thomas R. Kailbourn. *Pioneer Photographers from the Mississippi River to the Continental Divide: a Biographical Dictionary.* Stanford, CA: Stanford University Press, 2005.

Palmquist, Peter E., and Thomas R. Kailbourn. *Pioneer Photographers of the Far West: a Biographical Dictionary.* Stanford, CA: Stanford University Press, 2000.

Patera, Alan H., ed. *Your Obedient Servant: the Letters of Quincy A. Brooks, special agent of the Post Office Department, 1865–67.* Lake Oswego, OR: Raven Press, 1986.

Payne, Edward R. "Oregon Territorial Post Offices and Handstamped Postal Markings." *Oregon Historical Quarterly, Vol. 60, No. 4.* Portland: Oregon Historical Society, 1960.

Prosch, Thomas W. "Military Roads of Washington Territory." *Washington Historical Quarterly, Vol. 2, No. 2.* Seattle: Washington University, January 1908.

Ramsey, Guy Reed. *Postmarked Washington: Thurston County.* Olympia: Thurston County Historic Commission, 1988.

Rau, Weldon W. *Pioneering the Washington Territory: Reminiscences of Willis and Mary Ann Boatman, with a Review of Historical Events of Early Territorial Days.* Tacoma: Media Production Associates, 2003.

Reese, Gary Fuller. *A Documentary History of Fort Steilacoom, Washington, Second Edition.* Tacoma: Tacoma Public Library, 1984.

Robinson, Joan. "The Hard First Way Across the Mountains." *Columbia, The Magazine of Northwest History.* Tacoma: Washington State Historical Society, Summer 1988.

Sato, Mike. *The Price of Taming a River: The Decline of Puget Sound's Duwamish/Green Waterway.* Seattle: The Mountaineers, 1997.

Schoonmaker, J. M. et al. "Edward Jay Allen." *Circular No. 5, Series of 1916.* Philadelphia: Military Order of the Loyal Legion of the United States, Headquarters Commandery of the State of Pennsylvania, July 5, 1916.

Smith, Herndon. *Centralia: The First Fifty Years, 1845–1900.* Centralia, WA: The Daily Chronicle and F. H. Cole Printing Company, 1942.

Smith, James R. *San Francisco's Lost Landmarks.* Sanger, CA: Word Dancer Press, 2005.

Snowden, Clinton. *Rise and Progress of an American State.* New York: Century History Company, 1909.

Stevens, Isaac I. *Narrative and Final Report of Explorations For a Route For a Pacific Railroad near The Forty-Seventh and Forty Ninth Parallels of North Latitude from St. Paul to Puget Sound Washington Territory.* Washington, DC: 1860.

Swan, James Gilchrist. *The Northwest Coast.* Fairfield, WA: Ye Galleon Press, 1989.

Thurston County Pioneers Before 1870. Manuscript No. 134. Manuscripts compiled by Thurston County Historical Association and Thurston County Historic Commission. Olympia: Washington State Library, circa 1920. Also available on microfilm and Internet (see Online Resources).

Trosper, Don, and Janet Haag. *The History of Tumwater, Vol. 2, New Market.* Tumwater, WA: Tumwater Historical Association, 1987.

U.S. Army, Pennsylvania Infantry Regiment, 155th. *Under the Maltese Cross, Antietam to Appomattox, the loyal uprising in western Pennsylvania, 1861-65; campaigns 155th Pennsylvania, narrated by the rank and file.* Pittsburgh: The 155th Regimental Association, 1910.

Unruh, John D. Jr. *The Plains Across.* Chicago: University of Illinois Press: 1982.

Urrutia, Virginia. *They Came to Six Rivers: The Story of Cowlitz County.* Kelso, WA: Cowlitz County Historical Society, 1998.

"Van Ogle's Memory of Pioneer Days." *Washington Historical Quarterly, Vol. 8, No. 4.* Seattle: Washington University, October 1922.

Washington Territory Donation Land Claims. Seattle: Seattle Genealogical Society, 1980.

Wilkinson, George L. "Early Roads in the Fort Steilacoom Area." *Fort Steilacoom, Vol. X, No. 1.* Steilacoom, WA: Historic Fort Steilacoom, March 1993.

Winston, Dorothy. "Early Spanaway." Tacoma: Graphic Press, 1976.

Winther, Oscar. "Inland Transportation and Communication in Washington 1844-1859." *Pacific Northwest Quarterly, Vol. 30.* Seattle: University of Washington, January-December 1939.

Winthrop, Theodore. *The Canoe and the Saddle or Klalam and Klickatat.* Tacoma: John H. Williams, 1913.

Wright, E. W., ed. *Lewis and Dryden's Marine History of the Pacific Northwest.* Seattle: Superior Publishing Company, 1967.

Newspapers

Olympia had a succession of newspapers in its early days. First came the *Columbian* from September 11, 1852 until November 26, 1853. It was followed by the *Washington Pioneer*, which published from December 3, 1853 until January 28, 1854. Next came the *Pioneer and Democrat*, which operated from February 4, 1854 until May 31, 1861.

"About the Cascade Road." *Washington Pioneer*, December 24, 1853.

"Aged Settler of Coos Tells of Reaching Top of Mt. Adams in 1854." *The Portland Evening Telegram,* November 1909.

"An Old Pathfinder, Hon. Edward Jay Allen Returns to His Early Home—An Interesting Tale." *Yakima Herald*, August 15, 1889.

"An Old Pioneer and What He Says About the New Olympia." *Olympia Tribune*, June 17, 1891.

"Another Pioneer Gone." *Washington Standard,* July 7, 1893.

"Arrival of Immigrants." *Pioneer and Democrat,* September 16, 1854.

"Capt. McClellan." *Pioneer and Democrat,* March 11, 1854.

"Cascade Road Fund," *Pioneer and Democrat,* September 24, 1854.

"Col. Allen and His Indian Password." *Washington (D.C.) Herald,* November 5, 1913.

"Colonel and Mrs…" *Pittsburgh Post*, September 16, 1889.

Conover, C. T. "Just Cogitating: Edward Jay Allen Promoted Road Over Cascades." *Seattle Times*, June 10, 1956.

Conover, C. T. "Life of Dr. David Maynard, Colorful Pioneer." *Seattle Times*, April 5, 1951.

"Delinquent Tax List." *Steilacoom Puget Sound Herald,* March 4, 1859.

"Died." *Pioneer and Democrat*, October 30, 1857.

"Died." *Washington Standard*, February 23, 1861.
"Edward Jay Allen." *Chronicle Telegraph* (Pittsburgh), December 27, 1915.
Fleeson, Reese C. "Capt. Wm. H. Allen." *Pittsburg Dispatch*, February 1857.
"For Oregon." *Pittsburgh Daily Gazette and Advertiser*, April 2, 1851.
"From A Pioneer." *Morning Olympian*, December 4, 1898.
"Great Mass Meeting. The Peoples Road." *The Columbian*, October 8, 1853.
"James Longmire, Pioneer." *Tacoma Sunday Ledger*, August 21, 1892.
"Judge B. F. Yantis." Obituary. *Washington Standard*, February 15, 1879.
"Kansas Affairs." *Cleveland Herald*, June 20, 1856.
"Late and Interesting News." *Pioneer and Democrat*, January 6, 1855.
"Liberality," *The Columbian*, October 15, 1853.
"Lieut. Arnold." *Pioneer and Democrat*, May 20, 1854.
Light, E. A. "Incidents of the Early Days." *Tacoma Sunday Ledger*, June 19, 1892.
"List of Passengers." *Pioneer and Democrat*, December 23, 1854.
"Longmire Tells Naches History." *Yakima Herald*, March 12, 1923.
Mahlberg, Blanche Billings. "Letter Home From Olympia, O. T., 1852." *Tacoma News Tribune*, December 12, 1965.
Mahlberg, Blanche Billings. "Pioneer Trail Blazed by Edward J. Allen." *Tacoma Sunday Ledger-News Tribune*, September 6, 1953.
"Meeker Again Heeds Call of Old Oregon Trail." *Tacoma Ledger*, April 2, 1923.
"Messrs. Ensign, Kirtley and Blankenship." *Pioneer and Democrat*, August 19, 1854.
"Mr. Sargent has reached home again." *The Columbian*, October 22, 1853.
"Mr. Silas Galliher." *Pioneer and Democrat*, September 2, 1854.
"Olympia Bakery." Advertisement. *The Columbian*, March 26, 1853.
"On Thursday…" *Puget Sound Weekly Courier*, January 5, 1877.
"Oregon Alarmed." *The Columbian*, August 6, 1853.
"Passengers." *Daily Alta California*, February 17, 1855.
"Pioneer Brooks Dies at Port Townsend." *Morning Olympian*, July 8, 1908.
"Pioneer of State Passes Away In East." *Olympian*, January 9, 1916.
"Pioneer Steamers of Puget Sound." Unattributed clipping, Pamphlet Files, Washington State Library.
"Postscript. San Francisco Correspondence." (Sacramento) *Daily Union*, February 17, 1855.
"Proceedings on the Fourth." *The Columbian*, July 9, 1853.
"Proclamation." *Pioneer and Democrat*, March 4, 1854.
"Puget Sound Shipping Report." *Pioneer and Democrat*, January 27, 1855.
"Receipts." *Pioneer and Democrat*, September 30, 1854.
Richardson, David. "Friday Harbor Founder Had Colorful Life." *Seattle Times*, December 10, 1961.
"Road Across the Cascades." *Pioneer and Democrat*, July 8, 1854.
"Road Over the Cascade Mountains." *The Columbian*, April 30, 1853.
"Road to Walla Walla." *The Columbian*, July 16, 1853.
"Roll of Washington Pioneers." *Washington Standard*, November 1 and 8, 1889.
"Shipping." *Weekly Oregonian*, March 25, 1854.
"Shipping Intelligence." *Daily Alta California*, February 27, 1855.
"Swan." Obituary. *Morning Olympian*, February 20, 1904.
"Territory of Columbia." *The Columbian*, April 9, 1853.
"The Cascade Road." *The Columbian*, July 23, 1853.
"The Cascade Road." *The Columbian*, August 13, 1853.
"The Cascade Road." *The Columbian*, September 24, 1853.
"The Peoples Road." *The Columbian*, October 1, 1853.
"The Road Across the Cascade Mountains." *The Columbian*, May 28, 1853.
"To-Day!" *Washington Pioneer*, December 17, 1853.
"Tribute to Edward J. Allen, Esq." *The Columbian*, October 8, 1853.
"Visit to Mt. Rainier." *The Columbian*, September 18, 1852.
"Weed." Obituary. *Daily Olympian*, January 13, 1896.
"Whig Nominations." *Washington Pioneer*, December 31, 1853.
"Winthrop Allen, Civic Leader, Dies." *Louisville Times*, March 25, 1964.

Census and Genealogical Records

Donation Land Claim Notes, Washington. U.S. Bureau of Land Management, http://www.blm.gov/or/landrecords/survey. Also available as microform, Washington State Library, Northwest Micro 333.16.
United States Census, various years. Federal Census. Online. 2003 and updated 2007. Washington Secretary of State, http://www.digitalarchives.wa.gov.
Washington Territorial Census. Territorial/County Census. Online. Washington Secretary of State, http://www.digitalarchives.wa.gov.

Maps

"Edward J. Allen Addition to the City of Olympia" map. Plat Book, Vol. 8. City of Olympia Records.
Land Status and Cadastral Survey Records. U.S. Bureau of Land Management, http://www.blm.gov/or/landrecords/survey.
"Township 19 North, Range 2 West." *Thurston County, Washington map*. Unattributed. Washington State Library microform, Northwest Micro 912.7977.

Personal Correspondence

Clay-Poole, Scott, Ph.D. Personal correspondence relative to native plants of Washington. November 2009 and January 2013.
Dixon, Thomas W. Jr. Personal correspondence relative to railroads near Covington, Virginia. September 14, 2009.
Flora, Stephenie, compiler. "Isaac Smith, Pioneer of 1852." Personal correspondence via e-mail.

Online Resources

"An Act to Regulate Marriages." Laws of Washington Territory, 1854. Washington State Archives—Digital Archives. http://www.sos.wa.gov/history/images/publications/SL_washingtonlawsv1/SL_washingtonlawsv1.pdf#page=p=651
"Baron Karl von Reichenbach." *Gale Encyclopedia of Occultism & Parapsychology*. http://www.answers.com/topic/baron-karl-von-reichenbach.
Becker, Paula. "Saint Joseph's Mission at Ahtanum Creek…" HistoryLink.org. Essay 5825. http://www.historylink.org/index.cfm?DisplayPage=output.cfm&file_id=5285.
"Benjamin Penhallow Shillaber." *Chelsea Historical Society*. http://www.olgp.net/chs/people/ben.htm.
"Black, Samuel Watson." *The Political Graveyard*. http://politicalgraveyard.com/bio/black.html#565.92.76.

Blankenship, George E. "History of Olympia Lodge No. 1, F. & A. M. Olympia, Washington 1852-1935." *Olympia Historical Society*. http://olympiahistory.org/index.php?option=com_content&view=article&id=67:history-of-olympia-lodge-no-1-faam&catid=6:transcriptions&Itemid=2.

Blankenship, George E. "Lights and Shades of Pioneer Life, By a Native Son." *Olympia Historical Society.* http://olympiahistory.org/index.php?option=com_content&view=article&id=66:lights-and-shades-of-pioneer-life&catid=6:transcriptions&Itemid=2.

Blue, Fred. "Lane, Joseph (1801-1881)." http://www.oregonencyclopedia.org/entry/view/lane_joseph_1801_1881_/.

"Boat and Ship Types." www.abc.se/~pa/bld/shiptype.htm.

Brunelle, Jim. "The Maine Law." http://www.maine.gov/sos/kids/about/history.htm.

Buerge, David M. "Chief Seattle and Chief Joseph: From Indians to Icons." *University of Washington Libraries—Digital Collections*. http://content.lib.washington.edu/aipnw/buerge2.html.

"Charles C. Terry." *University of Washington Libraries—Digital Collections*. http://content.lib.washington.edu/cdm4/item_viewer.php?CISOROOT=/portraits&CISOPTR=199&CISOBOX=1&REC=1

"Churches of Washington." Access Genealogy. http://www.accessgenealogy.com/washington/churches_of_washington.htm.

"Claude-Étienne Minié." *Encyclopedia Britannica*. http://www.britannica.com/EBchecked/topic/384023/Claude-Etienne-Minie.

Crooks, Drew. "Intermarriage: The Example of John and Betsy Edgar." Leschi, Close Ties. Washington State Historical Society. http://stories.washingtonhistory.org/leschi/closeties/intermarriage.htm

"Emanuel Swedenborg." *Gale Encyclopedia of Occultism & Parapsychology*. http://www.answers.com/topic/emanuel-swedenborg.

"European and Euro-American History." Olympic National Park, History and Culture. http://www.nps.gov/olym/historyculture/upload/Euro-history.pdf.

"Fichtel Hills." *Encyclopedia Britannica*. http://www.britannica.com/EBchecked/topic/206007/Fichtel-Hills.

"François Norbert Blanchet." *Catholic Encyclopedia*. http://www.newadvent.org/cathen/02593a.htm.

"Free Soil Party." *Encyclopedia Britannica*. http://www.britannica.com/EBchecked/topic/218365/Free-Soil-Party.

Gaines, John Calvin III. "John Pollard Gaines." Salem Public Library. Salem Online History. http://www.salemhistory.net/people/john_pollard_gaines.htm.

Gough, Barry M. "New Caledonia." The Canadian Encyclopedia. http://thecanadianencyclopedia.com/articles/new-caledonia.

"Haida & Argillite." Simon Fraser University. www.sfu.ca/archaeology-old/museum/argillite/my_the1_home.html.

Journal of the U.S. House of Representatives, February 14, 1860, page 269. Library of Congress: A Century of Lawmaking for a New Nation. U.S. Congressional Documents and Debates, 1774-1875. http://memory.loc.gov/cgi-bin/query/r?ammem/hlaw@field(DOCID+@lit(hj05655)

Journal of the House of Representatives, December 13, 1860, page 73. Library of Congress: A Century of Lawmaking for a New Nation. U.S. Congressional Documents and Debates, 1774-1875. http://memory.loc.gov/cgi-bin/query/r?ammem/hlaw@field(DOCID+@lit(hj05710)

Lea-Scarlett, E. J. "Bochsa, Robert Nicholas Charles (1789-1856)." *Australian Dictionary of Biography*. http://adb.anu.edu.au/biography/bochsa-robert-nicholas-charles-3019.

McLagan, Elizabeth. "The Black Laws of Oregon, 1844-1857." BlackPast.org. http://www.blackpast.org/?q=perspectives/black-laws-oregon-1844-1857.

"Military Department, Indian War Muster Rolls, 1855-1856—Albert Eggers." Washington State Archives—Digital Archives. http://media.digitalarchives.wa.gov/WA.Media/jpeg/6AFA6B87F05CB9040BE56CF912C91DD9_1.jpg

"Military Department, Indian War Muster Rolls, 1855-1856—Charles W. Riley." Washington State Archives—Digital Archives. http://www.digitalarchives.wa.gov/Record/View/83B92E1AA18A71046B4145ABE026CB01

"Military Department, Indian War Muster Rolls, 1855-1856—Shirley Ensign." Washington State Archives—Digital Archives. http://www.digitalarchives.wa.gov/Record/View/E79B0FF6EAC105146D607F336922A28F

"Mounds of Mystery." Science Frontiers. http://www.science-frontiers.com/sf119/sf119p09.htm

"Mount Rainier." USGS Volcano Hazards Program. http://volcanoes.usgs.gov/volcanoes/mount_rainier/

Newell, Gordon. "So Fair A Dwelling Place." *Olympia Historical Society*. http://Olympiahistory.org/index.php?option=com_content&view=article&id=74:so-fair-a-dwelling-place&catid=6:transcriptions&Itemid=2.

Powell, Albrecht. "Pittsburgh – America's Most Misspelled City." http://pittsburgh.about.com/od/about_pittsburgh/a/spelling.htm.

Quinn, David B. "Madoc." *Dictionary of Canadian Biography Online*. http://www.biographi.ca/009004-119.01-e.php?&id_nbr=438&interval=25&&PHPSESSID=t8tt34o1l66b8k3fvrqi2njff6.

Rochester, Junius. "Maynard, Dr. David Swinson." HistoryLink.org. Essay 315. http://www.historylink.org/index.cfm?DisplayPage=output.cfm&File_Id=315.

"Samuel Augustus Mitchell." http://www.raremaps.com/makers/mitchell.html.

"Spiritualism." *Encyclopedia Britannica*. http://www.britannica.com/EBchecked/topic/560501/spiritualism.

"SS *Golden Age*." The Maritime Heritage Project. http://www.maritimeheritage.org/ships/index.html.

"SS *Uncle Sam*." The Maritime Heritage Project. http://www.maritimeheritage.org/ships/index.html.

"Steamboat Building in Elizabeth, PA." *Built at Elizabeth, PA*. http://freepages.history.rootsweb.ancestry.com/~jmohney/built_at_elizabeth,_pa.htm.

"The Adams Express Company." http://adamsexpress.com/files/u2/adams_history.pdf.

"Thomas Corwin." *Ohio History Central*. http://www.ohiohistorycentral.org/entry.php?rec=76.

"Thurston County Pioneers Before 1870." Washington Rural Heritage. http://www.washingtonruralheritage.org/cdm/search/collection/pioneers/field/all/mode/all/conn/and/order/title/ad/asc.

"Whig Party." *Encyclopedia Britannica*. http://www.britannica.com/EBchecked/topic/641788/Whig-Party.

"William Henry Aspinwall." The Panama Railroad. http://www.panamarailroad.org/aspinwall.html.

Wilma, David. "Stevens, Isaac Ingalls (1818-1862)." HistoryLink.org. Essay 5314. http://www.historylink.orgindex.cfm?DisplayPage=output.cfm&file_id=5314.

Index

Of Related Interest

Washington State

The Inaugural Decade, 1889–1899

Robert E. Ficken

"Residents of Washington, regardless of whether they lived east or west of the Cascades, agreed upon a fundamental point. Their recently-admitted state was, in every possible way, superior to southern California."

—Robert E. Ficken, in *Washington State*

In the sequel to his popular *Washington Territory*, author Robert E. Ficken deftly describes the turbulent first decade of statehood—ten years that laid the foundation for the following century and beyond.

"His best work to date."

—Carlos A. Schwantes

Photographs • notes • bibliography • index
6" x 9" • 304 pages (2007)
Paperback • 978-0-87422-288-3 • $21.95

Washington Territory

Robert E. Ficken

"This book will be invaluable to scholars of the Pacific Northwest for years to come."

—Daniel Herman, Central Washington University

Established in 1853, Washington remained a territory for thirty-six years. Statehood was only awarded after railroads finally unified the region in the mid-1880s. This comprehensive, one-volume chronicle is the definitive economic and political saga of territorial Washington.

Photographs • notes • bibliography • index
6" x 9" • 304 pages (2002)
Hardbound • 978-0-87422-249-4 • $35
Paperback • 978-0-87422-261-6 • $22.95

Shaper of Seattle

Reginald Heber Thomson's Pacific Northwest

William H. Wilson

"Looking at local surrounds, I felt that Seattle was in a pit, that to get anywhere we would be compelled to climb out of it if we could."

—Reginald Heber Thomson

During his tenure, city engineer Reginald Heber Thomson delivered a clean, reliable water supply, a workable sewage system, regraded streets, and more. *Shaper of Seattle* recounts the life and work of this extraordinary man and his devotion to the Emerald City.

"Wilson's book, generously illustrated with archival photos, brings the man and his legacy to life."

—Mike Dillon, *Queen Anne & Magnolia News*

Photographs • maps • notes • bibliography • index
8½" x 11" • 240 pages (2009)
Paperback • 978-0-87422-301-9 • $29.95